SCIENCE
BE DAMMED

How Ignoring Inconvenient Science
Drained the Colorado River

ERIC KUHN AND JOHN FLECK

**THE UNIVERSITY OF
ARIZONA PRESS**

TUCSON

The University of Arizona Press
www.uapress.arizona.edu

ISBN-13: 978-0-8165-4005-1 (cloth)
ISBN-13: 978-0-8165-4323-6 (paper)

Cover design by Carrie House, HOUSEdesign llc
Cover photo: *Delta of the Colorado River in the Grand Canyon* by mundoview/Shutterstock

The publisher and authors gratefully acknowledge the generous support for this book provided by the board of directors and staff of the Colorado River Water Conservation District.

The views and opinions expressed in the book are those of the authors alone and do not represent the views and policies of the District.

Library of Congress Cataloging-in-Publication Data
Names: Kuhn, Eric, author. | Fleck, John, 1959– author.
Title: Science be dammed : how ignoring inconvenient science drained the Colorado River / Eric Kuhn and John Fleck.
Description: Tucson : The University of Arizona Press, 2019. | Includes bibliographical references and index.
Identifiers: LCCN 2019009202 | ISBN 9780816540051 (cloth)
Subjects: LCSH: Water resources development—Government policy—Colorado River Watershed (Colo.-Mexico) | Water-supply—Government policy—Colorado River Watershed (Colo.-Mexico) | Science—Political aspects—United States. | Science and state—United States.
Classification: LCC HD1695.C6 K85 2019 | DDC 333.91/62097913—dc23
LC record available at https://lccn.loc.gov/2019009202

Printed in the United States of America
♾ This paper meets the requirements of ANSI/NISO Z39.48–1992 (Permanence of Paper).

CONTENTS

SCIENCE BE DAMMED

Introduction and Purpose

In the epilogue to his classic *Water and the West*, historian Norris Hundley captured the problem at the heart of sustainably managing the waters of the Colorado River Basin in the twenty-first century:

> "Basin of Contention" would be an apt name for what generations have called the Colorado River Basin. A limited supply of water in a vast and semi-arid region is hardly a recipe for tranquility among those who covet the water. The drafters of the compact were clearly aware of that truism, but they nonetheless failed to determine with reasonable accuracy the long-term annual flow of the Colorado River.[1]

The implications of that failure are profound. There is less water in the Colorado River, over the long run, than the farms and cities across and adjacent to the Colorado River Basin have come to expect and depend on. One of us, Eric Kuhn, now the retired general manager of the Colorado River Water Conservation District, spent a career wrestling with the implications of that failure. The River District spans the Colorado River headwaters of western Colorado. The struggle for sustainability starts there, because about two-thirds of the river's waters start there. During the many plane flights and hotel evenings spent traveling the basin for gatherings of water managers wrestling with the problem, Hundley's book was often in Eric's backpack as he thought about the question of how we got to this point.

Two things came to mind. The first was an observation by his predecessor and mentor at the River District, Rolly Fischer (1928–2016). One of Fischer's favorite sayings about the river was that the tried-and-true method to solve disputes in the Colorado River Basin was to promise the combatants more water than was available in the river, then hope a future generation would fix the mess. The second was the words of Hundley. His book, first published in 1975, is the go-to resource on the negotiations and ratification of the Colorado River Compact. By the time the book's second edition was published in 2009, the problem of scarcity, of the overallocation of the Colorado River, had become acute, and Hundley reflected on the reasons. When the framers of the 1922 Colorado River initially parceled out the water, he wrote,

> they had a glaring need for sound information, but no concerted attempt was made to call on the scientific community for help. The drafters were mesmerized by their desire for haste and their political and personal goals. Without authoritative data, they had an opportunity to pick and choose information that best suited their interests and uncertainties. And that is what they did. The situation would not significantly change until others recognized and studied the importance of tree-ring data—data that revealed a distinct pattern, going back centuries, of severe and lengthy droughts, and the probability this pattern will continue in the future. The consequences of the compact remain with us.[2]

While Kuhn was toting Hundley to Colorado River Basin management meetings, the other of us, John Fleck, was trying to make sense of the implications of another classic in the Colorado River literature. Purchased from a rare book dealer in the winter of 2012, Fleck's copy of hydrologist E. C. LaRue's 1916 U.S. Geological Survey report entitled *Colorado River and Its Utilization* bears the yellowing of a century. A New Mexico–based journalist drifting away from newspaper work toward an academic career, Fleck was puzzling over the same questions as Kuhn. As Albuquerque, his hometown, increasingly came to depend on Colorado River water, overallocation and scarcity made the problems of the basin personal. LaRue's work seemed to hold the key to understanding how science had informed the decisions that brought us to this point.

The conventional story of how this happened, enshrined in Marc Reisner's seminal 1986 book *Cadillac Desert*, is that in overestimating the available supply of water, the authors of the 1922 Colorado River Compact did the best with what they had—"about eighteen years of streamflow measurement. . . . During all of that period, the river had gone on a binge."[3] Our great misfortune, that conventional story would have us believe, was that the compact's framers could not have known that they were allocating the water during unusually wet times.

Hundley's criticism brings us part way to correcting the mistake in that conventional story. Yes, as Reisner argues, the 1922 Colorado River Compact negotiators proceeded without a good working knowledge of the river's hydrology. Yes, as Hundley argues, the commissioners were profoundly incurious about the question. But the history is more complicated.

The decision-makers actually had available, had they chosen to use it, a relatively thorough, complete, and almost modern picture of the river's hydrology. Had they chosen to use the information, they had at hand more than Reisner's "eighteen years of streamflow measurement." They had available, but chose not to use, data suggesting a much smaller river in the years prior to the eighteen years on which they were relying. Had they taken the science seriously, they almost certainly would have concluded that the Colorado River had less water than the common assumptions underpinning their race to develop the river.

We cannot know what they might have done had they used this information. Accepting this science might have left them with a flow too low, and projected drought periods too severe to reach the compromises necessary to carve up the river among the states and develop its waters. The basic problem was neither the lack of good science nor the ability of the decision-makers to understand the basin's hydrology. The decision-makers were in fact intelligent, accomplished, and skilled professionals. The problem was that in an era driven by politics of competition for a limited supply of river water and federal dollars, the decision-makers had the opportunity to selectively use the available science as a tool to sell their projects and vision for the river's future to Congress and the general public. This approach used by the compact commissioners set a precedent that would continue for decades and color most of the major policy decisions on the river in a way that still wields undue influence today.

Today, the Colorado River is fully used, most would say overused. Unless it is by careful design, as happened in the spring of 2014 with an experimental environmental "pulse flow," not a drop of Colorado River water makes it to the Sea of Cortez. Demands for the river's waters already exceed the available supply, a situation that will only grow more difficult with continued growth and the impacts of climate change.

Science offers new tools to respond to the challenges. Advancements in hydrology and climatology, including the reconstruction of river flows and climatic conditions from past centuries, and a longer and more complete gauge record, give us far more hydrologic and climate information than was available to our predecessors. However, because of this wealth of technical data and science and the nature of the human decision-making process, water agencies still have ample opportunity and political incentive to continue to justify critical decisions by selectively ignoring science and technical information that is inconvenient to their needs and interests.

We need a realistic assessment of the misuse of science in the past, to understand the way those mistakes have become embedded in the rules used to govern the river today, and a realistic assessment of the variability and uncertainty in the river's supply of water going forward. All of that must feed into the development of a stable system of governance that recognizes that demand now exceeds supply on the Colorado River.

We need to understand how we got here if we are to make better decisions in the future.

CHAPTER 1

The Colorado, a River in Deficit

The boat ramp at Las Vegas Bay, once a shimmering recreation mecca on the shores of Lake Mead, now ends in a row of concrete barricades and desert sand. A short hike through the scrub leads to an incongruous flowing river, the effluent from the Las Vegas metro area's wastewater treatment plants, flowing the last few miles to Lake Mead.

The floating marina that once anchored Las Vegas Bay here was moved in 2002, towed to deeper water as Lake Mead declined.[1] The great reservoirs integrate the Colorado River's two stories—nature's water flowing in, and humans taking it out. Too little of the first, or too much of the second, is in the long run unsustainable. At the bottom of the old Las Vegas Bay boat ramp, you can look up and see which version of the story is playing out etched in the hillsides above, old shorelines long since left dry by Lake Mead's decline.

At the 2013 meeting of the Colorado River Water Users Association, Arizona water manager Tom McCann gave the reservoir's problem a name. Hundreds of people had gathered in Caesar's Palace twenty miles west of Las Vegas Bay for the annual meeting of the Colorado River water management community. It was a tense affair. Lake Mead is the nation's largest reservoir, the anchor of a hydraulic empire built over the twentieth century that spans nine states in two nations. In the previous year, it had dropped more than

Figure 1 The abandoned Las Vegas Bay. Photo by John Fleck.

thirteen feet—enough water to meet Las Vegas's needs for nearly six years. Bureau of Reclamation program manager Carly Jerla warned the audience that within the next few years there was a chance Mead would not have enough water to meet the downstream users' needs.

There is a tendency for water managers to blame drought and climate change when reservoirs drop and water becomes scarce, but McCann asked the audience in the big Caesar's Palace ballroom to confront a more uncomfortable reality. Even without drought and climate change, which by 2013 were clearly taking their toll on the Colorado River, there had never been enough water to meet the long-term promises the river's governing bodies had made to the people who had come to depend on the Colorado's water. There was, to use a phrase that would come to dominate Colorado River discussions in the years that followed, a "structural deficit" on the river. By that McCann meant that even under normal water supply conditions, the rules created by the region's political leaders over the previous century had allocated more water on paper than the river could supply in reality. This was not an aberration based on unusual climate. This was inevitable.

The open and public discussion launched in 2013, that farms and cities in the United States and Mexico were overusing the available water supply, was startling to some and refreshing to others. To those in charge of the river's water, it felt unique. For much of the previous hundred years, the river's managers had inhabited a fantasy world in which abundant supplies would make all things possible. The disconnect between fantasy and the river's real-world hydrology, which had been increasingly clear in private conversations and obscure technical reports, was now being laid out in full public view. But a closer reading of the river's history reminds us that what McCann was saying was not as new as it felt like that December morning in Las Vegas. McCann's message of a river in deficit had been delivered nearly a century ago by some of the leading scientists studying the river. It had been repeated many times since then. Again and again, science suggested less water in the river. It was ignored as promoters and politicians jockeyed to grab their slice of the water. They built dam after dam, with canals to sluice off ever more water from the shrinking river and its dwindling reservoirs. Today, the projects that resulted from all of that jockeying are in place, along with cities and farms depending on the water's flow. But there is not enough water to fill them. Some of the problem can be attributed to climate change, which is drying the river. But even absent climate change, we would be in trouble. The twenty-first century's problems on the river are the inevitable result of critical decisions made by water managers and politicians who ignored the science available at the time they were being made.

In 1925, U.S. Geological Survey scientist Eugene Clyde LaRue tallied what was known at the time about the river's flow, potentially irrigable farmland, and growing cities, and concluded that if we built the dams and canals to use all the water being allocated on paper in the 1920s to meet all the anticipated demand, the Colorado River would be in deficit.[2] Two contemporary scientific analyses also done in the 1920s, one by LaRue's U.S. Geological Survey colleague Herman Stabler and one by a board of experts commissioned by Congress and led by retired Gen. William Sibert, backed him up. Despite the warnings from LaRue, Stabler, and Sibert, Congress approved the Colorado

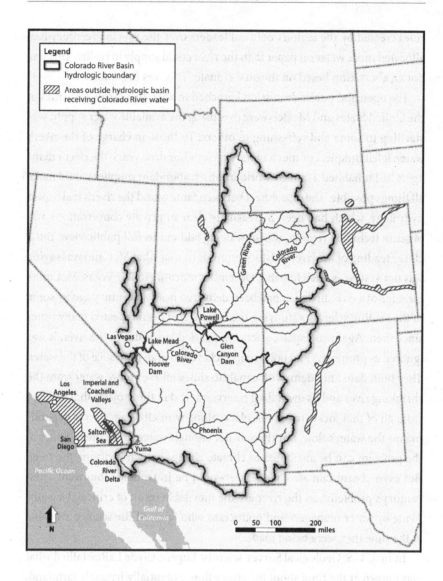

River Compact, the foundation of the legal edifice that has come to be called the "Law of the River" and launched a century of dam building that allowed the sparsely populated deserts of the Colorado River Basin and surrounding areas to bloom. By ignoring the best available science of their day, they set in motion decades of decisions that would end in the overuse seen today.

The plane flight from Las Vegas to Denver, across the canyon country of the Colorado Plateau and the high Rockies that generate the river's water, is stunningly beautiful. The flight starts in the Colorado River's Lower Basin, the arid country of Nevada, Arizona, and Southern California that has come to depend on the water of the Colorado River. The typical flight path leaving Las Vegas's McCarran Airport toward the east climbs out of the Las Vegas Valley over the Black and Boulder Canyons, carved over millions of years by the Colorado River. Out the plane window the most notable feature today is Hoover Dam, tiny from this vantage point, and the vast expanse of Lake Mead piled up behind it. It shimmers blue in the late afternoon sun, but a passenger in a window seat can't help but notice its white "bathtub ring," the layer of minerals left behind around its rocky desert rim as the reservoir drops.

In December 2013, as McCann spoke in nearby Las Vegas, Lake Mead was less than half full and dropping fast. The bathtub ring had been growing since the early years of the twenty-first century. The years 2000–2004 were exceptional, the driest five years in a century of detailed records of the Colorado River's flow. But they were simply a prelude to what was to come. After a seesaw from moderately dry to a single wet year in 2011, 2012–13 followed as one of the driest two-year periods in the river's modern history. But water users kept using as much as ever, and storage in the river's two big reservoirs, Lake Mead and Lake Powell, continued their decline.

By the time it reaches the ocean, the Colorado River has drained approximately 242,000 square miles of the southwestern United States and another 2,000 square miles in Mexico. As it makes its 1,450-mile course to the ocean it flows through four different major landscapes: the Rocky Mountains; the Colorado Plateau; the Basin and Range Province; and its own delta. Over its course to the ocean, it drops from over 14,000 feet on its highest peaks to sea level. The growing season varies from three months in high mountain valleys to seven months at 5,000 feet in the lower valleys of Colorado to year-round by the time it leaves the Grand Canyon.

Although one of the great rivers of North America, the Colorado's flow as measured by the natural discharge at its mouth is modest. Both the Colorado River and the Columbia River Basin drain an area of approximately 250,000 square miles, but the average annual natural discharge of the Columbia is thirteen times that of the Colorado. When compared with other U.S. rivers based on natural discharge, the Colorado is slightly larger than the Hudson, and about the same as the Illinois River.[3] If anything, though, the Colorado's small size relative to its basin makes its role *more* important. The Hudson River Valley is awash in water. If you need water in this part of the world, the Colorado is your only source.

The river's geography poses a second great challenge. The majority of its flow originates in the high country in its upper reaches, while the greatest chances to use its water—its largest cities and most productive farmland—are in its lower reaches.

A flight from Denver to Las Vegas crosses the boundary between those two very different geographies, following the course of the Colorado River over the Grand Canyon, past a historic northern Arizona river crossing called Lees Ferry.

A passenger flying east over the canyon country that defines this part of the Colorado Plateau—a flight both authors have made many times, peering out the windows to make sense of the landscape—could see beyond Lees Ferry the waters of Lake Powell, the river's second great reservoir. Completed in the mid 1960s, Powell sits astride the Arizona-Utah border, a gleaming pool of blue in the late afternoon sun but another casualty of lower flows in the river. In 2013, holding 10.8 million acre-feet, Lake Powell was just 44 percent full. The Colorado River's flow that year was less than 60 percent of the long-term average.

The divide between Lower and Upper Basin is crucial to understanding the history of the development of the Colorado River. Much of its basin is dry, and the river's water is central to the lives of the forty million people who live in and around its basin and use its water. But while most of the water use happens in the Lower Basin, most of the water itself comes from the Upper Basin, falling as winter snows in the high mountains and filling the river with the melt of spring and summer. The tension over this fact—water coming from one place and being used in another—is central to the river's story.

It was here, as the river flowed through the canyon country, that LaRue in 1923 gathered some of the data on the Colorado River's flow that proved so prescient, and that was so thoroughly ignored nearly a century ago. A photo of the hydrologist shows him shirtless in the summer sun measuring the flow of Nankoweap Creek in the Grand Canyon, part of an expedition to fill in our understanding of how much water the Colorado River had and how we might use it. LaRue had been studying the river's hydrology for a decade, but the report that resulted from the 1923 expedition became the definitive account, in its day, of how much water the Colorado had to offer.

When LaRue and his colleagues emerged from the canyon country at Needles, California, at the conclusion of their 1923 expedition, his initial explanation of what he and his colleagues had found was brief. "We were sent to secure certain information for the government," he told the *Los Angeles Times*, "and we got it."[4] At that point in the early history of the Colorado River's development, the river shortfall we see before us now was nothing more than shadows of the future cast in the pages of the analyses LaRue and his colleagues began producing.

But what he began reporting in the following two years bears a striking resemblance to the "structural deficit" McCann laid out before the Colorado River Water Users Association nearly a century later. The numbers were of necessity imprecise when LaRue published them in the 1920s, but the bottom line was the same as McCann's in 2013. What LaRue identified as inevitable growth of water use in the river's upper reaches, especially in the state of Colorado, would reduce its flow before it reached the desert canyon country. Downstream, vast areas of desert landscape awaited the water that would turn them into bountiful farms. Cities to the west, in Southern California, were outstripping their local supplies and looking for more. "If the population of this region continues to increase," LaRue wrote, "a new source of domestic water must be found." That "new source" was a canal through the desert to the Colorado River, one of many such schemes of the region's boosters to use Colorado River water to overcome the arid land's shortcomings. Basing their plans on the flow during an unusually wet few decades, the boosters thought they could pull it off. But LaRue's math, taking into account past droughts the boosters were unwilling to consider, suggested there was not enough water to meet their aspirations.[5]

The list of people who followed LaRue with a similar message is long: Herman Stabler, who had accompanied LaRue and colleagues on their 1923 expedition and published his independent analysis in 1924; William Sibert, drafted by Congress in 1928 to provide an independent assessment; Reclamation commissioner Harry Bashore in 1945, as Congress was considering a treaty allocation a share of the river to Mexico; Northcutt Ely in 1946, who took the stage at a meeting of the same Colorado River Water Users Association that McCann addressed sixty-seven years later, warning of an inevitable shortfall; Royce Tipton, who in 1965 warned of future shortages because of the river's inability to meet the paper apportionments in the Colorado River Compact; Arizona's G. E. P. Smith, who in the 1920s and again in the 1940s cautioned his state that the river's yield was far less than what the decision-makers were claiming.

The difference in 2013 was that McCann was no longer describing a future possibility. LaRue, Stabler, Sibert, and those who followed were dismissed in their day as hypothetical pessimists. But as McCann spoke, the white bathtub ring circling Lake Mead could no longer be ignored.

From today's perspective, the obvious question is why projects kept being built, and increasing amounts of water diverted, in the face of the evidence being presented. Our answer is that like the approval in 1928 of the Boulder Canyon Project Act which ignored the findings of LaRue's 1925 report and the Sibert board report, the fate of projects at each step of the way were decided by the politics of the day, not science. The concept of "entitlement" ruled the actions of the states. The original founding documents of the river's management were seen to create a promise of water, and few were willing to question whether the Colorado River could deliver.

Terminology Guide for Readers

Like most professions, those that are involved in the management and use of water in the West have their own somewhat unique terminology. For readers who do not routinely use "water talk," some of the key terms you will encounter in this book are defined below:

An **acre-foot** is the volume of water that would cover an acre of land with water one foot deep, or 325,853.3 gallons. The typical suburban home in the Denver area consumes about half an acre-foot per year, but actual uses vary considerably primarily depending the lawn size. The productive farmlands near Yuma consume almost 5 acre-feet per acre per year. Streamflow is measured in **cubic feet per second**, or "cfs." It also is common to measure the annual flow of streams in **acre-feet per year**. (Most of the rest of the world uses the metric terms *cubic meter* for a volume of water and *cubic meters per second* for flow.)[6] The **natural flow** of a stream is the estimated flow that would have been in the stream absent man-made uses. The estimated natural flow of the Colorado River at its **mouth** (the very bottom of a river—an ocean, terminal lake, or confluence with a bigger stream) is about 16 million acre-feet per year. This may seem like a lot of water, but compared to other well-known U.S. rivers, like the Mississippi, over 500 million acre-feet per year, or the Columbia, over 200 million acre-feet per year, it is quite modest.

This book is primarily about the development and use of the **surface water** available from the Colorado River. In contrast, **groundwater** lies in underground aquifers and is pumped to the surface with wells. There is a complex interaction between ground and surface water. In many cases, wells are used to intercept groundwater that is tributary to the adjacent stream and, thus, groundwater use often impacts the surface water in the stream.

Dams store water in **reservoirs** for later use during droughts or when the normal flow of the stream is insufficient to meet downstream demands. Most major dams are built for multiple purposes, including supplying downstream uses, flood control, and hydroelectric power. Diversion dams are used to divert surface water into supply diches, canals, and aqueducts.

A **water right** is a legal right to the use of water recognized by a governmental entity, normally the state. In most states, a water right is perfected by putting water to a **beneficial use**, such as the irrigation of crops; the use of water for domestic, industrial, or municipal purposes; for generating hydroelectric power; and now, in many states, for instream purposes, such as fish flows and recreation. Water rights are documented in a variety of ways, depending on the state, such as court decrees, permits from state agencies, or contracts. Water rights can be for a given flow, in cfs, or an amount, in acre-feet. The amount of **consumptive use** of a water right is often different than the amount **diverted**. For example, an irrigator might divert 10 cfs to a field, but if 5 cfs returns to the stream as runoff from the field, only 5 cfs is consumed. As a rule of thumb, most irrigation uses are about 50 percent consumptive, whereas municipal indoor uses are about 10 percent consumptive. Exports (or **transbasin diversions**) out of a river basin are 100 percent consumptive (to the basin of origin) and instream and hydroelectric power generation uses 0 percent (fully nonconsumptive).

In most western states, the doctrine of **prior appropriation** controls surface water use. Water rights are given a priority date based on when the water was first put to beneficial use. Water uses are queued up for the available supply in order of their priority. In most of the rest of the United States, **riparian doctrine** controls water use. A riparian right gives a landowner the right to the reasonable use of water from a stream flowing through or bordering property subject to that use not impacting other riparian uses on the stream (where there is plenty of water, this is not a problem).

A **compact** is a negotiated contract among two or more states approved by the U.S. Congress and the legislatures of the participating states. The use of compacts to divide up the uses of water on interstate streams is common. In situations where there is no compact, the U.S. Supreme Court has settled disputes among states. Where a compact does exist, the Supreme Court has often had interpret or resolve disputes among compact parties.

The **Colorado River Compact** is an agreement among the seven Colorado River Basin states and the United States that was negotiated in 1922 and approved by Congress in 1928. The major provisions of the Colorado River Compact are summarized as follows:

Article I defines the purposes of the compact.

Article II defines the key terms used in the compact, including defining the **Upper Basin** as that portion of the Colorado River Basin that drains into the river above **Lee Ferry**, a point on the river one mile downstream of the confluence of the Colorado River and the Paria River, and the **Lower Basin**, which is that part of the basin that drains into the river below Lee Ferry. The **States of the Upper Division** are Colorado, New Mexico, Utah, and Wyoming, commonly called the Upper Basin states. The **States of the Lower Division** are Arizona, California, and Nevada, commonly called the Lower Basin states. Note that three states—Arizona, New Mexico, and Utah—have lands in both basins.

Article III deals with the allocation and use of the waters of the Colorado River. **Article III(a)** apportions, in perpetuity, to each the Upper and Lower Basins, 7.5 million acre-feet per year of beneficial consumptive use. **Article III(b)** allows the Lower Basin to increase its beneficial consumptive use by an additional million acre-feet per year. **Article III(c)** defines the obligations of each basin to Mexico in the event of a future treaty (which happened in 1944). **Article III(d)** requires the States of the Upper Division to not deplete the flow of the river at Lee Ferry below 75 million acre-feet every consecutive ten years. **Article III(f)** and **(g)** provide for a future apportionment after October 1, 1963 (which never happened).

Article IV declares the Colorado River nonnavigable and makes the use of water for hydroelectric power generation subservient to the use of water for domestic and agricultural purposes.

Article V provides for cooperation among the basin state and federal officials for determining facts such as river flows, consumption, and the use of water.

Article VI addresses how controversies between or among the compact states should be handled (we don't think it has ever been used).

Article VII provides that the compact shall not be construed as affecting the obligations of the United States to American Indian tribes.

Article VIII protects rights that were perfected at the time the compact was signed and removes the threat that rights in the Lower Basin could affect rights in the Upper Basin once 5 million acre-feet of storage is constructed (it has).

Article IX provides that nothing in the compact prevents a state from taking legal action to protect rights or enforce the compact.

Article X provides that the compact can be terminated with the unanimous agreement of the states, but that any right established while the compact was in place would remain unimpaired.

Article XI provides that the compact does not become effective until ratified by each of the states and U.S. Congress. Note that this provision was waived by all the states, except Arizona, after Arizona first refused to ratify the compact.

Lee Ferry, **Lees Ferry**, and **Lee's Ferry** are often used interchangeably but describe two different places. The Colorado River Compact defines Lee Ferry as "a point in the main stream of the Colorado River one mile below the mouth of the Paria River." Under the compact, Lee Ferry is the dividing point between the Upper and Lower Basins and the point on the river where there are prescribed flow requirements. Lees (or Lee's) Ferry is the name of the historic settlement located at the mouth of the Paria River where John D. Lee once operated a ferry. While commonly called Lee's Ferry, the official government name for the historic site has no apostrophe. For this book, the authors have chosen to use Lee Ferry when referring to compact-related subjects and flows.

CHAPTER 2

The Origin of Conflict

People had long lived in the Colorado River Basin when we first began applying what we might call "modern science" to its management. Fitting to its time, that first foray came in the form of a military expedition, as the U.S. Army sent a young lieutenant named Joseph Christmas Ives to explore and report back on what he found. Ambitious and resourceful, Ives was more concerned with the river's navigability than the potential use of its water. Turning the river into the sort of navigable waterway that opened the U.S. interior to commerce in the pre-railroad era could be done, Ives concluded, but it would be difficult. There was simply too little water.[1]

For much of history, the question of how much water there was in the Colorado River as a whole was unimportant. The native communities in the basin before European immigrants arrived, along with the early settlers from the south and east, developed water at a local scale, never using enough to have a significant impact on the basin as a whole. Small private irrigation systems began in the late 1800s. This began to change in the early 1900s, when large diversion systems and supply canals, built primarily with federal government help, enabled a rapid expansion of irrigation in the Imperial Valley, central Arizona, the Yuma area, and the mid-elevation valleys on the upper river. The region's new residents were removing water from the Colorado at a scale that was beginning to matter for the river as a whole.

After World War I, needs for flood control, water security for the Imperial Valley, farming opportunities for war veterans, and hydroelectric power to support rapid population growth in the Southern California coastal plain focused attention on the development of the Colorado River. It also scared the upper river states, afraid that a rapidly growing California would suck up all the river's water before they had a chance to develop their shares. Legal uncertainties and competition for the river's resources resulted in a regional call for a peace on the river that would foster further regional development, and demonstrated the need for science to answer the seemingly simple questions—how much water does the Colorado River have and how much of it is available for future development?

Native people have lived in the Colorado River Basin for thousands of years. Pueblo communities at times flourished and at times were disrupted by the vagaries of the basin's climate. Irrigation by sophisticated canal systems was common in the Gila and Salt River Valleys. But lacking the ability to move large amounts of water or food over long distances, they were constrained by local climate and hydrology. Remains of old agricultural developments are found throughout the valleys of Central Arizona, a geo-engineering so sophisticated that modern irrigation canals often follow the routes pioneered by the valley's native residents.[2]

As our young nation expanded westward the Colorado River Basin was among the last regions of the continental United States to be developed. By the early to mid-1800s, traders, trappers, and explorers traversed the Basin. The discovery of gold in California in 1849 brought more settlers into the lower river and the Mormon migration to Utah brought settlers into the Green River Basin.[3]

The early stirrings of the application of scientific tools to understanding the basin can be traced to the mid-1800s. In 1851, the U.S. Army established Fort Yuma. That first step prompted the War Department in 1857 to send Ives up the Colorado River to determine how far inland river navigation was possible. He made it to about the present location of Hoover Dam.

Lt. Ives documented the farming done by native communities and considered the possibility of large-scale irrigation in the Lower Colorado River Basin, but only briefly, concluding that it would be impractical because of the risks posed by the river's annual spring and summer floods.

It was a prescient insight, but it went no further than the musings of the young lieutenant. In his report, Ives famously concluded this about the deserts of the Lower Colorado River: "The region last explored is, of course, altogether valueless. It can be approached only from the South, and after entering it, there is nothing to do but leave. Ours was the first, and doubtless will be the last, party of whites to visit this profitless locality. It seems intended by nature that the Colorado River along the greater portion of its lone and majestic way shall be forever unvisited and unmolested."[4]

In the summer of 1875, U.S. Army Lt. Eric Bergland and his team, attached to the famed Wheeler Survey, traveled east by mule train in the heat of the desert summer to gather information about the region and, they hoped, answer the question of whether some portion of the land might be irrigable with water from the Colorado River. Bergland made the first crude measurement of the flow near the site of Hoover Dam. His measurement, 18,410 cfs, which would translate to 13.3 million acre-feet per year, and his estimate that 3.7 million acres of land could be irrigated are strikingly close to today's values. However, he readily acknowledged that the harsh desert was unlikely to come into agricultural development any time soon. The most important contribution of Bergland and his supervisor, First Lt. George M. Wheeler, was their call for better data on how much water was available for use as settlers moved west. "The measurement of the waters of this stream . . . is a matter of importance," they wrote, "and . . . should be ordered at an early day."[5] It would be another four decades before their call was taken up in earnest.

It was in the upper reaches of the Colorado River that the first water development by settlers from the east began. In 1854, irrigation began from Black Forks, a tributary to the Green River near what is now the Wyoming-Utah border. After the Civil War a mining boom in the central Rockies and Arizona, cattle ranching, and the construction of transcontinental railroads all led to additional development within the basin. In the 1860s and 1870s, early settlers began irrigating in Colorado's high mountain valleys, growing food for the miners and their pack animals. In 1877, Thomas Blythe filed for rights to Colorado River water in California, the first foray into large-scale irrigation in the desert valleys of the Lower Colorado River.

After the native Indians were largely driven from the region in the 1880s, settlers moved into the mid-elevation deserts in western Colorado, forming ditch companies to tap the rivers for irrigation water. They built gravity canals and ditches in the broad shale valleys surrounding Grand Junction and Montrose and the U-shaped higher elevation glacial valleys. By the early 1900s, much of the region's bottomlands adjacent to the Colorado River and its principal tributaries were under irrigation for farming and ranching. Water development was growing to a scale that had impacts no longer entirely local.

This called for action at a scale not before attempted in the United States. John Wesley Powell, in his seminal 1878 *Report on the Lands of the Arid Region*, described the challenge as an integration between the science of data gathering and analysis—the work of the great surveys—combined with the engineering needed to bring about the arid lands' "redemption," and importantly "the legislative action necessary to inaugurate the enterprises by which these lands may eventually be rescued from their present worthless state."[6]

The science in which Powell was engaged, and which his successors pursued in the development of the Colorado River, was not science in its purest sense, the pursuit of truth for its own sake. It was science of the sort described by Powell colleague and contemporary W. J. McGee when he wrote that "the course of nature has come to be investigated in order that it may be re-directed along lines contributing to human welfare. . . . Now is the time of conquest over nature."[7]

The clash between the honesty of science and the politics of the day was there from the beginning. In the late 1800s, a nationwide economic depression caused a crash in mineral prices, and serious drought brought pressure on the U.S. Congress to come to the aide of local irrigators in the West. It also brought impatient congressmen and senators into conflict with science.

In 1890, Powell, then director of the nation's new science agency, the U.S. Geological Survey (USGS), was locked in battle with Senator William Stewart of Nevada. Powell wanted money from Congress to complete a detailed irrigation survey that had begun at Stewart's urging in 1988. Powell envisioned a comprehensive survey that would be used to identify the best lands to irrigate, and to locate and plan for dams and supply canals. By 1890, Senator Stewart and his western colleagues had become impatient with the expense and length of time to conduct the survey. They wanted more immediate action. In what would become the model for the future, Powell lost.[8] In the conflict between political expediency and need for science to better inform decision-makers, science loses, and in the long run it most often leads to bad decisions and future trouble.

In 1902, Congress passed the Newlands Act (named after Nevada Senator Francis G. Newlands), also known as the Reclamation Act.[9] The Reclamation Act along with the Homestead Act and other federal statutes were designed to encourage the settlement of the arid western United States by the "reclamation" of land for farming. The theory behind the legislation was that if the federal government assisted local irrigation by providing lands for new irrigation and water for to irrigate those lands, reclamation would pay for itself. The Colorado River Basin was an immediate beneficiary of the law. Among the early projects built by the Reclamation Service were the Uncompahgre Project near Montrose and Delta, Colorado; the Grand Valley Irrigation Project near Grand Junction, Colorado; the Strawberry Valley Project in Utah; the Yuma Project in Arizona; and the Salt River Project in central Arizona.

The idea that irrigation could pay for itself also brought private developers to the region. In southeastern California, the Colorado River created the Imperial Valley through the deposition of trillions of tons of sediment in the rift created by the separation of the Pacific Plate from the North American Plate.[10] Most of the valley north of the U.S.-Mexico border is below sea level. The valley soil is rich and fertile, and the growing season is year-round, but without water it was a fruitless desert. In the late 1890s, a private land developer, the California Development Company, identified a gravity route and built a canal to deliver water from the Colorado River near Yuma to the valley. They called it "the Alamo Canal." The first Colorado River water arrived to eager valley settlers in June 1901. It was a historic vanguard of one of the largest modern irrigation projects in the United States.

The availability of irrigation water triggered rapid settlement. By 1904, there were over fifteen thousand settlers, seven hundred miles of canal, and seventy-five thousand Imperial Valley acres under irrigation.[11] In 1905, poor design and cheap construction of the civil works that diverted water out of the river into the Alamo Canal led to disaster.[12] The Colorado, swollen from

Gila River floodwaters, broke through the Alamo Canal diversion head-works, and the entire Colorado River flowed into the Imperial Valley for sixteen months. This flood inundated thirty thousand acres of irrigated land and created what is now called the Salton Sea. Controlling the flood required the help of the Southern Pacific Railroad and bankrupted the California Development Company. The railroad ended up owning the companies' water delivery assets. In response, in 1911 the valley farmers organized the Imperial Irrigation District (IID) to purchase the irrigation assets back from the railroad and run the irrigation system as a public entity.

The change was crucial. Formation of the district gave the farmers a public entity that could serve as a political voice for the valley, to contract with the Reclamation Service for an All-American Canal, and lobby for the national intervention needed to bring a flood control dam to the river. Since the Imperial Valley was below sea level and down gradient from the Colorado River, the risk of floods remained a major concern of the valley and would be the major motivation for the construction of upstream storage. The legal arrangements that IID's predecessor made to build and maintain the canal in Mexico gave Mexican farmers the right to use half of the water IID was diverting, but the cost of the construction and maintenance of expensive flood levies was the U.S. farmers' burden. A canal route entirely within the United States would solve the problem of sharing water with Mexico, but it wouldn't solve the flooding problem. To solve their problems, the valley's residents needed to engage in the politics and policy of water management at the regional, national, and international levels.

Meanwhile, rapid growth in the Los Angeles area was creating a demand for both water and electric power. With the construction of the Owens Valley Aqueduct, Los Angeles had already shown water could be moved hundreds of miles from where it was available to where it was needed. Private power companies were now exploring the canyons of the Colorado River looking for dam sites to supply future customers.[13]

The success of irrigation in the Imperial Valley and the need to protect it, irrigation development and rapid settlement of the areas served by the Reclamation Service projects, and the need for water and power in Southern California meant that no longer could water management there be a purely local affair.

Legal and Political Developments
in the Upper Basin

In the early 1900s, interstate water law governing the movement of water across state lines was new. One of the early legal scholars and leaders was Colorado's Delph Carpenter, who became the father of the interstate water compacts.[14]

To understand interstate water law, we first need to look at water law within the individual Colorado River Basin states. In six of the seven basin states, surface water is strictly allocated through "prior appropriation." Prior appropriation is based on the concept of "first in time is first in line." It is a system of queuing up users for a limited supply. A user establishes an appropriative right by putting the water to "beneficial use." California operates a merged system. However, for those water users in California that use Colorado River water, prior appropriation controls.

Early on, Carpenter and others viewed states as sovereigns that could fully use all waters originating within or flowing into a state with no regard or legal commitment to downstream states. The Supreme Court thought otherwise. In the 1907 case of *Kansas v. Colorado*, the court ruled that in the case of interstate streams, states are "coequals," and thus they must share the resource in an "equitable" manner.[15] The 1922 *Wyoming v. Colorado* case involved water rights on the Laramie River, a tributary of the North Platte River that originates in the mountains west of Fort Collins, Colorado, and flows north into Wyoming. In the early 1900s, a Colorado developer proposed diverting water from the Laramie River into the adjacent South Platte basin for irrigation. Diverting this much water from a relatively small stream would have had major impacts on existing users downstream in Wyoming. In 1911, Wyoming went to the Supreme Court to protect its users. Ultimately the court ruled that since both Colorado and Wyoming were prior appropriation states and given the 1907 decision, an "equitable" settlement of the dispute was to apply prior appropriation to the Laramie River as a whole.[16] This decision protected the earlier, or prior, water rights in Wyoming, but rendered the Colorado project infeasible.[17]

Carpenter was involved in the Laramie case and understood that Colorado was at risk of losing. So, even before *Wyoming v. Colorado* was finally decided in June 1922, he feared that if the lower Colorado River states (Arizona and

California) were to use additional water as the result of projects built by the Reclamation Service, these states would claim a legal priority requiring delivery of a certain quantity of water. Without a compact to protect their ability to develop at a pace consistent with their own needs, the upper river states could grow their Colorado River use only at the mercy of the lower river. Like the Laramie River project, future Colorado River projects in the upper river could be rendered infeasible by downstream senior projects.[18]

A compact, essentially a negotiated agreement among the states, setting out allocations in advance rather than leaving them to prior appropriation, could "prevent a free-for-all race to see who develops the fastest because it would assure each participant state that its rights were permanently protected no matter how long it might take to get its economic engines running," historian Daniel Tyler explained. Further, Tyler wrote, a compact would "avoid costly litigation, assure the supremacy of equitable apportionment instead of prior appropriation across state lines, eliminate future embargoes by the Reclamation Service, and settle title to water rights on the rivers before the construction of dams and reservoirs."[19]

It was not just the development of irrigation projects that concerned Carpenter. He also feared the development of large hydroelectric power dams on the lower river without an agreement protecting future upper river supplies. Big power-generating dams might lock in big supplies of water for the communities that came to depend on them.[20]

Thus, in 1919, 1920, and again in 1922 when California legislators introduced bills in Congress for federal assistance to build a canal entirely within the United States to supply water for the Imperial Valley and a large dam and reservoir on the lower river, representatives from the upper river states were opposed absent an interstate agreement protecting their states' future water use.[21] The controversy over the legislation showed that what was needed was a social contract between the faster- and slower-growing states. If the faster-growing states could agree to leave some water in the river for the future needs of the slower-growing states, then all of the states could work together for federal legislation that would develop the river.

Just as importantly, the political pressure on all of the basin to reach a settlement on the Colorado River was huge. In addition to the flooding, water supply and electrical power generation concerns in the lower river and the legal concerns in the upper river, there was pressure from business, community, and political leaders who believed the comprehensive development of the river for irrigation, power and industrial purposes would be the catalyst for a broader economic development of the entire southwestern United States.

Colorado's Carpenter first championed the concept of a legal compact or contract among the states at a regional meeting in Denver in August 1920. It quickly caught on and received backing from all seven state governors.[22] The next step was to gain congressional approval.

On August 19, 1921, Congress passed legislation giving its approval to an interstate compact on the Colorado River and authorizing federal participation in the negotiation of the compact. The legislation required that the compact negotiations be completed by the end of calendar year 1922.

Early Understanding of the Hydrology

Prior to 1922, the primary sources of information on the hydrology of the Colorado River and its tributaries were water supply papers prepared by the U.S. Geological Survey. The USGS began installing river gauges in the basin in the late 1800s. The first was on the Gila River at Buttes in 1889, installed when the USGS was under the leadership of John Wesley Powell. Powell brought to the agency a grand vision of the role of science in informing development of the arid West, a project he referred to as "redemption" of the arid land.[23] In the Upper Basin, the first gauge was installed in 1894 on the Green River near Green River, Utah. The USGS published data from its gauging stations in annual water supply papers.

In the late 1890s, the USGS began making investigations of irrigation projects, water storage opportunities, and flooding problems in the basin. Most of these studies were also published as water supply papers. In 1897, Arthur Powell Davis, nephew of John Wesley Powell, authored a study of irrigation near Phoenix, Arizona.[24] Davis would later become chief engineer and director of the Reclamation Service and would be a major figure in negotiations of the Colorado River Compact. Working for Davis, J. B. Lippincott is often given credit for first identifying the Boulder and Black Canyon dam sites in his report on the Colorado River in 1904.[25]

Our modern understanding of the Colorado River dates to the 1916 publication of Water Supply Paper 395, *Colorado River and Its Utilization*, by
Eugene Clyde (E. C.) LaRue of the USGS—the first comprehensive engineering report on the Colorado River Basin. Water Supply Paper 395 represented
the first great fork in the road in the development of the water supplies of the
basin. LaRue laid out a road map for the development of the river, the dams
and diversions needed to fully exploit the river's water in the development of
the economy of a twentieth-century West. But he also offered a caution—the
river's water "is not sufficient to irrigate all the irrigable lands lying within
the basin," LaRue concluded.[26]

His report described the river and its major tributaries, summarized and
evaluated the available hydrology records, inventoried existing irrigation
systems, identified potential irrigation opportunities, and identified and
evaluated reservoir sites for river control, irrigation supply and hydroelectric
power purposes. As we shall see in later chapters, LaRue would continue to
be a major and controversial engineering and hydrology contributor to the
technical understanding of the river system.

Water Supply Paper 395 marked the first attempt to think about the river
in its entirety, "so that a broad view of the possible utilization of the whole
river could be obtained," as USGS chief hydrologist Nathan Grover, LaRue's
boss, wrote in the report's introduction.

In Water Supply Paper 395, LaRue estimated from the few gauges available
at the time the historical flow of the Colorado River at a number of important locations, including the Green River at five locations, the Colorado River
above the Green (then referred to as the "Grand"), the Colorado River below
the Green and the Grand, and the Colorado River at Laguna Dam on the
Arizona-California border, near where the Colorado River leaves the United
States and enters Mexico.[27] LaRue's estimated average annual flow of the Colorado River at Laguna Dam for the period of 1895 to 1914 was 16.2 million
acre-feet per year. LaRue made no attempt to determine what the "natural
flows" might have been, the amount of water that would have been in the
river absent dams and diversions upstream of the measurement point.[28]

When LaRue published his work, Lee Ferry, now so important to Colorado River measurement and governance, was just a dot on the map near the
location of a river crossing referred to as Lees Ferry.[29] LaRue had no reason

to calculate the flow of the river at this location. LaRue estimated the flow of the Colorado River at several upstream points as a percentage of the flow at Laguna Dam. One of those locations is the Colorado River at the mouth of the Paria River, which is close to the location now referred to as Lee Ferry. If natural flow had been considered an important number in 1916, LaRue would have concluded that the natural flow at Lee Ferry for 1895–1914 was about 16.2 million acre-feet per year.[30] Over time, this would become one of the most important measurement points, and one of the most important numbers, on the Colorado River.[31]

LaRue's discussion of potential reservoir sites is thorough and complex. He calculated how much storage upstream would be needed to manage floods and provide a steady flow on the lower river for irrigation. His answer was 18 million acre-feet. He identified a number of dam and reservoir sites that were subsequently developed, including what are now the Flaming Gorge, McPhee, Taylor Park, Navajo and Davis Dams. Interestingly, LaRue made no mention of the Glen Canyon Dam site or either the Boulder Canyon or Black Canyon sites in his discussion of river regulation.[32] However, he did mention Boulder Canyon as a potential hydroelectric dam site.

The First "Paleo" Reconstruction

In Water Supply Paper 395, LaRue also made the first crude effort at what is today called a "paleoclimate reconstruction"—using indirect evidence of past climate to estimate the river's flow in the time before the introduction of stream gauges. LaRue used historical levels of the Great Salt Lake, which rises and falls in step with greater or lesser inflow from the snow-capped mountains that surround it. Inflows to the lake can be estimated by the records of lake elevation. A reliable lake elevation gauge goes back to 1875. LaRue supplemented the gauge using the memories of pioneers back to 1850. He examined the correlation between inflows to the lake and Colorado River flows and found a good relationship between them. Initially, he limited his analysis to conclusions concerning periods of high, low, and average runoff. He would later use this method to reconstruct annual flows at Lee Ferry. It was the first step toward an analysis that within a decade would provide, before a compact would be signed and Hoover Dam authorized, a warning that deep and long-lasting droughts were a central feature of the hydrology of the Colorado River.

Reclamation Seizes an Opportunity

Initially the Reclamation Service, created by the 1902 Reclamation Act, was a branch of the USGS. In 1907, it became a separate service. In theory, this separated the science and data agency (USGS), from the on-the-ground engineering agency (Reclamation). In the early years, this distinction was blurred, and it set up a competition between the agencies. Later, when Reclamation became a promoter of the projects it wanted to build, it had an incentive to control the science and data, and it did so.

In 1914, Arthur Powell Davis became director of the Reclamation Service and that agency's efforts to study comprehensive approaches to the development of the Colorado River accelerated. With Davis, IID found an eager ally to solve its flooding and water security problems. Addressing the problem of delivering water to the Imperial Valley without going through Mexico led to the formation of the All-American Canal Board in February 1918, created by an agreement between the IID board and the secretary of the interior. The formation of the board would set in motion two forces that would shape the development on the Colorado River. The formation of the board was the first small step toward an interstate compact among the states and the federal government on the Colorado River. Second, the agreement set in place the alliance between a federal agency, the Reclamation Service, a water delivery agency (IID) that needed access to the resources of the federal government, and the elected officials that both wanted to meet the needs of their constituents and controlled the Reclamation Service's purse strings.

The board issued its report in July 1919. It concluded that an All-American Canal, bypassing the Alamo Canal in Mexico and delivering water directly to Imperial entirely within the United States, was feasible and recommended its approval by Congress and immediate construction. It also recommended that "the United States should undertake the early construction of storage reservoirs on the drainage basin of the Colorado River as part of a comprehensive plan for the betterment of the water-supply conditions throughout the entire basin."[33] The report makes the case for storage for both flood control and water supply concluding that "whenever the river drops to a low stage early in the season as was the case in 1915 and in 1918, the demand of the irrigator upon the lower river will be in excess of the water supply. This situation will become more pronounced when all the land in the Imperial Irrigation District and in the Yuma project susceptible of cultivation shall have been brought under irrigation." However, the report does not specifically recommend the construction of the Boulder Canyon Dam or any other storage project by name.

Other than to formally recognize the reality that during low flow conditions, irrigation demands on the lower river would exceed supply without upstream storage, the All-American Board report included only a limited discussion of the hydrology of the river system. It included only one table of river flows at the Yuma gauge and a few other upstream gauges for the period of 1895–1918.[34] Under the discussion of the silt problem, the report states "the river's annual discharge may be placed at about 17,000,000 acre-feet." It simply presumed there would be enough water. One of the report's authors was Dr. Elwood Mead, then working for the California Land Settlement Board. For Mead, who would later become Reclamation commissioner during the design and construction of Hoover Dam, it was his first major effort on the Colorado River.

The All-American Board report led to the first of many major congressional actions regarding the Colorado River. The 1920 Kinkaid Act authorized the preparation of a comprehensive report on the Colorado River by the Reclamation Service. It would be formally known as *Problems of the Imperial Valley and Vicinity*. The report would be commonly referred to as the "Fall-Davis" report, named after Arthur Powell Davis and Albert Fall, the Interior secretary when it was issued. Although the report would not be formally completed until February 1922, much of the information included in the report was public before the compact negotiations began in January 1922 and thus was available to the negotiators. A preliminary report had been issued in November 1920.[35]

The Fall-Davis report incorporated and expanded on LaRue's work. In the letter of transmittal, Director Davis acknowledges the contributions of

E. C. LaRue and Water Supply Paper 395. The report's recommendation for the authorization and construction of the Boulder Canyon Project would become one of the major milestones in the development of the Colorado River and, as we shall see in the next chapter, a major motivation for the basin states to complete a compact. By 1920, the engineering and technology of concrete dams had advanced to the point that a high dam in Boulder or Black Canyon was considered feasible. The full proceedings of a conference dedicated to the engineering and construction of the Boulder Canyon Dam, dated December 12, 1921, would be appended as a formal section of the Fall-Davis report.[36]

A dam of this size and magnitude would not only capture the attention and imagination of the region, but ultimately the entire nation. The Reclamation Service used the Fall-Davis report to make the case that Boulder Canyon (or its Black Canyon alternative) was its preferred storage site. The report includes a comparison with the Glen Canyon Dam site. LaRue and others considered Glen Canyon a better alternative for the first major dam on the river.[37] The report makes the argument that while Glen Canyon is a good site for water storage, Boulder Canyon is preferred because it is in a better location to control floods on the lower river and, since "any large reservoir on the Colorado must depend for its financial feasibility upon the availability of an adequate market for not less than half a million horsepower of electric energy within economical transmission distance," Boulder Dam, being much closer to Southern California, was much preferred.[38]

While the Fall-Davis report would significantly advance the future development of the river, its treatment of the hydrology of the Colorado River was confusing. The report includes two important tables that, as we shall see, would be misunderstood and misused throughout the compact negotiations. The first of these tables, "No. 1—*Average discharges of principal tributaries*," shows the flows and drainage areas for the river and its major tributaries: the Green, Grand, San Juan, Gila, and "Other areas except Gila." There is no explanation as to the sources of the data or the period or periods of record used. Nor is there any discussion of the methodology used to develop the numbers.

The report's second important table is "No. 6—*Discharge of Colorado at Laguna Dam*."[39] Laguna Dam is a few miles upstream of the confluence with the Gila River. At that time Laguna Dam was the diversion point for the Yuma Irrigation Project, one of the earliest Reclamation Service projects on the Colorado River. There is no actual gauge, so the flows were estimated based on other gauges. Again, the sources of the data are not stated in the report. The Fall-Davis report annual flows at Laguna Dam are similar, but not identical, to flows at Laguna Dam shown by LaRue in Water Supply Paper 395.

Table 6 shows the average annual flow at Laguna for 1899–1920 as 16.4 million acre-feet per year, close to LaRue's average of 16.2 million acre-feet. It was not adjusted to account for the changes in river flow over time caused by growing agricultural use in the basin. In 1899, depletions upstream of Laguna were about 750,000 acre-feet per year. By 1920, they had increased to between 2.5 million and 3 million acre-feet per year. This confusion between actual flows in the river, natural flows, what they would have been absent upstream water use, and flows adjusted to a common depletion level showed up repeatedly in the early use of hydrology to underpin Colorado River decisions.

Together, Water Supply Paper 395, the All-American Canal board report, and the Fall-Davis report became the first steps toward answering the basic questions of how much water flowed in the Colorado River and how much was available for future development. They would be interpreted by Arthur Powell Davis, his Bureau of Reclamation colleagues, and state officials to conclude that the Colorado River could provide a water supply of at least 20 million acre-feet per year. What we now understand is that this period was not only wet, but exceptionally wet—one of the three or four wettest twenty- to thirty-year periods in the last 1,300 years.[40]

Although the Fall-Davis report and Water Supply Paper 395 showed similar river flows over the period of 1899–1920, there were also important differences. First, LaRue cautioned that due to growing irrigation uses upstream of Laguna Dam, the average flow at that point was misleading. Second, LaRue understood that a more complete picture of the water supply required going back before 1900 to look at a longer period of record. And, third, he made a fundamentally different conclusion concerning the adequacy of the available water supply to meet future demands. His conclusion that the water supply was not sufficient to meet future needs put him at odds with Arthur Powell

Davis, Davis's Reclamation colleagues, and most of the basin's water officials and politicians.

Could the Fall-Davis report have presented a more complete picture of the basin's water supply? Yes. For use in planning future projects, LaRue noted that gauged flow records need to be corrected for the growth in basin-wide irrigation. The authors of appendix B of the Fall-Davis report understood this concept and used it for their estimate of the future water available at the Boulder Canyon Dam site.[41] Since the flows shown at Yuma and Laguna were never corrected in a similar manner, they left an unrealistically high picture of the water available for the river basin as a whole.

Further, since drought conditions in the 1880s and 1890s were well documented and, in part, responsible for the Reclamation Act, it is a mystery why senior interior and state officials did not ask for a specific study of the water available for future use. Perhaps since their goal was to reach political agreements and promote the development of the river and since the Fall-Davis report suggested there was sufficient water, there was no incentive to do so. This was not science being used to provide the best possible understanding of the river. This, for the first time, was science being misused in support of a political path the boosters had already chosen.

The first gauge was not installed at Lees Ferry until the summer of 1921. Water year 1922 is the first full year for which an actual measured flow at Lee Ferry is available. Ironically, the gauge was installed not at the urging of the Reclamation Service or the states. It was accomplished by E. C. LaRue with a significant funding contribution from Southern California Edison, a power company interested in the development of hydroelectric power in the Glen Canyon region.

CHAPTER 3

The Colorado River Compact

By 1920, E. C. LaRue was growing concerned about the path of Colorado River development. The difference between his cautionary conclusion in Water Supply Paper 395 about the sufficiency of the Colorado River's flow and the optimism of the Reclamation Service's Arthur Powell Davis lurked in the background of discussions that seemed to be leading in what LaRue viewed as the wrong direction.

LaRue had offered, in his Water Supply Paper 395, a careful analysis of the river's flow and the potential demands for its water. Based on his understanding of the relatively few available gauge measurements, the limited period of record, and his concern with the uncertainties concerning past droughts, he concluded that, without a basin-wide plan, there would not be enough water to meet the aspirations of the seven states. Davis, in the Reclamation Service's 1919 annual report, had confidently reported "sufficient water supply in the Colorado River . . . to supply all future irrigation requirements."[1] LaRue saw that as a recipe for disaster. If full development of all the Colorado River Basin's irrigable land was pursued, "there will be a serious water shortage," LaRue wrote in an April 7, 1920, letter to his supervisor, the U.S. Geological Survey's Nathan Grover. LaRue suggested a meeting between the technical experts at the two sister agencies, his own USGS and Davis's Reclamation Service.[2] The meeting seems never to have taken place, but if it did, it had no effect. The development juggernaut, fueled by the optimism of Arthur Powell Davis, surged forward.

As government officials in the winter of 1921–22 prepared to begin negotiating a Colorado River Compact, LaRue tried again. This time he took the unusual step for a government employee of stepping outside his institutional chain of command and writing directly to Herbert Hoover, then secretary of commerce and, more importantly, likely to be chair of the soon-to-be convened Colorado River Compact Commission. LaRue offered his expertise to help compact negotiators understand the hydrology of the Colorado River. Hoover was so uninterested in LaRue's offer that he left a response to his assistant, Clarence Stetson, who in a January 4, 1922, letter offered a polite "thank you, we'll be in touch." Before the Compact Commission even began its meetings, the path had been chosen. The optimism of Arthur Powell Davis and the Fall-Davis report, rather than the caution of E. C. LaRue, would guide the negotiation of the Colorado River Compact and become embedded in a century of development and conflict.

Converging on Washington, D.C., in late January 1922, the representatives of the seven U.S. Colorado River Basin states had other things on their minds than the question LaRue was trying to raise. Dividing up the Colorado River's water would be hard enough without the nagging question of whether there was enough water to do what they hoped to do. No one had ever attempted anything like what they were attempting, and the question of how to do it was entangled in deep, unresolved questions.

- How would they prevent the fast-growing states like California from using the law of prior appropriation to stake an early claim to the Colorado River to the detriment of the slower-growing states?
- How much of the water originating in a state could be held for that state's use, and how much would have to be shared with others downstream?
- How would an agreement among the states be monitored and enforced?

At the center of these questions was the forty-four-year-old lawyer from Greeley, Colorado, Delph Carpenter. In late January 1922, as Carpenter boarded a train in Greeley, Colorado, for the two-day trip to Washington, D.C., he was at the top of his game. Carpenter was one of the nation's most distinguished lawyers. That Colorado governor Oliver Shoup would call on him to represent Colorado's interests on the Colorado River Commission was never in question. Carpenter held and championed two water law principles as sacrosanct. The first was that Colorado as a sovereign power had complete control over the water resources that originated in or flowed through its territory. The second was his unquestioned belief in the wisdom of Colorado's cherished prior appropriation doctrine. Yet as Carpenter made his way east, he knew problems were on the horizon. It was that pesky Wyoming case.[3]

As special counsel for Colorado, Carpenter was charged with defending Colorado's right to divert water from the Laramie River.[4] But downstream in Wyoming, farmers were already diverting water from the Laramie, and they had been there first. They had, in the language of western water law, undisputed "senior rights." After two oral arguments before the Supreme Court, Carpenter now feared the unimaginable. The U.S. Supreme Court, primarily a bunch of easterners, might grasp and apply the logic of prior appropriation on an interstate basis. At that point, his two principles could no longer coexist. Prior appropriation would be the torpedo that sank state sovereignty and, as a relatively slow-growing state, Colorado would be in trouble. This was about more than the Laramie, a relatively small river on Colorado's northern border. This implicated the Colorado, the state's largest river, its namesake. There, the state of Colorado would be at the mercy of the faster-growing and more economically powerful downstream states, especially California. When Carpenter arrived at Washington D.C.'s Union Station, he had devised an approach that, if successful, would protect his principles, but he knew it would not be an easy sell, and his window to sell it before the court ruled would be short.

1922, the Year of Negotiations

After Congress authorized a compact in 1921, President Warren Harding appointed Herbert Hoover as the United States commissioner. Hoover, a young, well-educated, articulate, world-famous engineer from California, was an ideal choice.[5] Harding's appointment of a high-profile cabinet officer to the commission was seen by the basin states as a sign that the president fully supported an agreement among the states. Hoover's principal technical advisor was Arthur Powell Davis, director of the Reclamation Service and author of numerous technical reports on the Colorado River Basin, including the Fall-Davis report.[6]

Gathering in Washington, D.C., in January 1922, the commissioners' first action was to elect Hoover chairman. In his greeting, Hoover laid out the goals he hoped a compact might achieve. From the beginning, the primary goal was development. The presumption was that there was plenty of water to do it. All that was needed was a deal to enable construction of dams and to decide who got how much. "It is hoped that such an agreement . . . will prevent endless litigation which will inevitably arise in the conflict of states' rights, with delays and costs that will be imposed on our citizens through such conflicts. . . . The problem is not as simple as might appear on the surface for while there is possibly ample water in the river for all purposes if adequate storage be undertaken, there is not a sufficient supply of water to meet all claims unless there is some definite program of water conservation. The Commission will, therefore, inevitably be driven into the consideration of a program looking further than the immediate legalistic relationship of the states if it is to find a solution."[7]

To understand the river's hydrology, the commission first turned to Davis, who wasted no time in telling them what they wanted to hear—with enough reservoir storage to capture water during wet years to be used during dry years ("conservation," in the language of the day, preventing it from being "wasted" to the sea), the Colorado River had enough water to do everything they would ask of it. "With proper and sufficient conservation," Davis explained, "there would be sufficient water for the irrigation of all the lands that could be favorably reached from the standpoint of economics within or adjacent to the Colorado Basin."[8]

Davis cautioned that "investigations of the Basin are no means complete" and sketched out the important development differences and challenges in the river's upper and lower reaches—"the preponderance of water in the Upper Basin and the preponderance of land in the lower basin and the difficulty of development in the Upper Basin." But the basic hydrologic principal had been established and would not be further questioned—there was enough water for all.

The conclusion that there was sufficient Colorado River water for all reasonable needs, or as he referred to it, land that could be "favorably reached," was a common theme of the Reclamation Service under Davis. It was the clear message from the Fall-Davis report. He and Hoover hoped that if there was enough water for all such irrigable lands in the basin, this would provide a simple formula for apportioning water among the states.[9] Tally up the available acres, allocate enough water to irrigate them all, and you would have a reasonable foundation for a deal. Davis knew that if that message could be accepted by the basin, it would remove the litigation threats and open the door for development that his agency would lead. What Davis failed to recognize was that too many of the states were interested in obtaining far more than just enough water to meet their reasonable needs.

The first attempt to reach a compact under this simple premise occurred at the commission's sixth meeting on January 30, when two subcommittees, one on water requirements and one on water supply, reported back to the full commission.

Arthur Powell Davis and the water requirements subcommittee used data on existing and future irrigation uses from both the federal government, the Reclamation Service and USGS, and from the individual states. Both federal agencies had recently completed surveys of irrigable lands.[10] State estimates

Table 1 Current and projected use (in acre-feet per year)

	1920 use	Reclamation projections for future use	State projections for future use
Upper Basin	2,267,000	3,991,500	8,099,500
Lower Basin	3,530,700	4,149,400	4,432,000
Total United States	5,797,700	8,140,900	12,531,500
Mexico	836,000	2,674,000	2,480,000
System total	6,633,700	10,814,900	15,011,500

Source: Authors' summary of tables A and B, Colorado River Commission, *Minutes and Record of the First Eighteen Sessions*, sixth meeting.

were based more on the wishes and aspirations of the states than on-the-ground information.[11]

The conflict between science and the politics of river development was now engaged. In offering projections far above Reclamation's, the states were engaged in high-stakes negotiations. If irrigable acreage was to be the standard upon which the Colorado River was to be divided, at least some of the states reckoned they must have *a lot* of irrigable acreage, regardless of what the engineers and surveyors had concluded. The major surprise was the magnitude of the difference in the upper river states of Wyoming, Colorado, Utah, and New Mexico. Representatives of some states, notably Colorado's Carpenter, were defensive about the difference between their numbers and Reclamation's. The Colorado numbers, Carpenter told the other commissioners, were the result of "careful analysis."[12] Others admitted the paucity of data on which they were basing their claims. Utah's R. E. Caldwell frankly acknowledged the data were "inadequate." Collectively, the Upper Basin states' estimated needs exceeded the Reclamation Service estimate by over 4 million acre-feet of water per year. In contrast, the Lower Basin states offered a more realistic perspective, only 283,000 acre-feet higher than the Reclamation estimate.[13]

How Much Water Is Available for Allocation?

While they were deliberate and strategic in their discussion of the numbers associated with future water demand, the compact commissioners and their technical advisors were strikingly cavalier about the crucial question of how much flow the Colorado River had, and therefore how much water they had available for allocation. As a result, this first great discussion of the allocation of the river's water embodied, in one place, all the confusions and mistakes that would dog efforts to manage the Colorado's flows in the years to come.

It is unfair to criticize the fact that they began the process with misunderstandings. They were doing something entirely new, with no base of experience to guide the kind of decisions they needed to make in allocating the Colorado River's waters. Misunderstandings were to be expected. What *is* fair to criticize is their fundamental lack of curiosity about the issues, given that there were experts available—most notably E. C. LaRue and his USGS colleagues—who *had* thought carefully about the problems at hand, who understood the pitfalls hidden within the numbers they hoped to use. Each state brought to the process both a principal negotiator and an engineering advisor who was or should have been expert in the technical details of water management. Rather than using that expertise, the negotiators took at face value a set of numbers based on a jumble of confusions and discrepancies—about where and when water was being measured, and about what those measurements might tell them about how much water the Colorado River might be able to provide to meet their desires for future development.

Instead of asking those questions they took at face value numbers contained in the Fall-Davis report, something never intended to serve as the hydrologic basis for a water allocation agreement. They appear to have done so out of expedience. The Fall-Davis report numbers seemed to be telling them they had enough water to do what they wanted to do, and they saw no advantage in asking too many questions about whether the numbers were right.

The details of their mistakes—how they used the available flow data, what was being measured at the critical gauge locations and how their mistaken understanding influenced their decisions—are complex. But those details are crucial to understanding what followed—how a mistaken understanding of the amount of water available in the Colorado River became deeply embedded in the decisions that shaped the modern West.

Today, when we analyze projects or agreements that use or impact Colorado River water, we try to start with a common set of flow data. Normally we start with the river's "natural flow" at Lee Ferry and other critical points on the river. By "natural flow," we mean a carefully calculated estimate of what would have passed that point on the river absent upstream dams and diversions. Naturalizing the flow in that way allows us to separate out and judge the impact of specific proposed actions, allowing us to make more careful decisions.

In 1922, the compact commissioners had no such agreed-upon standard. But they did have data. The Fall-Davis report provided the foundation for the compact negotiators' understanding of the river's available supply in the form of flow data at three locations near the downstream end of the

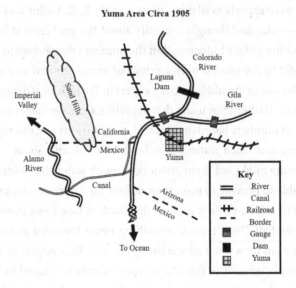

Yuma Area Circa 1905

Colorado River—Yuma (below the confluence with the Gila), Laguna Dam (above the confluence), and the Gila River near its mouth, at Dome, Arizona.

The complex geography of the river basin—where water was being added to the river, and where it was being removed, as it flowed past the three measurements points—led to confusions which caused a significant overestimate of the available supply of water.

The gauge at Yuma, which provided the underpinning for the primary Fall-Davis estimate of the river's flow, had been in place since 1902. The second point on the river for which both the Fall-Davis report and LaRue's Water Supply Paper 395 reported annual and average flows was at Laguna Dam, located a few miles upstream of Yuma and the diversion point for the Yuma Project. There was no river gauge at Laguna Dam. Before 1902, the flows were estimated based on only a handful of upstream gauges. Beginning in 1903, data from a gauge on the Gila River at Dome, installed in 1903, was subtracted from the gauged flow at Yuma.[14]

The confusions about place were compounded by confusions about time—over what period of record had the measurements been taken? The time period matters for two important reasons. The first involves the hints first offered by E. C. LaRue (later reinforced by Herman Stabler and others) that droughts might have reduced the river's flow for significant periods of time in the decades before the Yuma gauge was installed in 1902. The second problem with the period of record was more subtle, involving the significant differences between actual flow and the "natural flow"—what would have been there absent upstream use. The confusion meant that the commissioners would use the average flows at Yuma, Laguna Dam, and Dome to make important policy decisions. In doing so, they failed to understand the impact of the annual growth in irrigation use both on the Gila and Colorado Rivers upstream of the gauges.

It's clear the commissioners and their advisors were confused about what exactly was being measured at the different gauge locations, the policy implications of using average flows, and the problems with the different periods of record presented in the Fall-Davis report. That all this should have been confusing to them is to be expected. But their sin, which became the original sin of Colorado River Basin management, was a lack of humility in the face of their ignorance. Because while uncertainties where unavoidable given the modest nature of the available data, they had at their disposal a team of federal scientists, led by E. C. LaRue, who had carefully thought through the implications of the uncertainties. The commissioners never asked for their help.

The commissioners first assumed that the water available to them to meet the additional irrigation needs of the states and Mexico was 17.3 million acre-feet per year, the actual average flow of the Yuma gauge from 1902 to 1920.[15] The choice of the Yuma gauge without careful analysis to fully understand what was being measured was a terrible decision. The Yuma gauge is located below the Gila River, but above the diversion point where water, about 2.7 million acre-feet per year in 1920, was being diverted for delivery to Mexico and the Imperial Valley. The commissioners had a lengthy discussion of the location of the Yuma gauge during the sixth meeting, but it didn't clear up the confusion.[16]

The most serious error may have been the failure to take upstream uses into consideration. Water use above the Yuma gauge had been growing between 1902 and 1920, meaning that a single average flow over the entire period was sure to mislead. In 1902, the level of irrigation upstream of the Yuma gauge was about 1 million acre-feet per year. By 1920, that had risen to about 4 million acre-feet per year. Thus, flow at the Yuma gauge in the years before 1920 would have been lower had there been a 1920 level of development consistently through the period. Failing to take this into account threw off the numbers, making it seem as though the commission had more water to hand out than the actual flow of the river would support. To use as a water availability metric to meet future irrigation uses on new lands, the entire record should have been corrected to 1920 levels of development.[17] Based on information available at the time, correcting the Yuma gauge flows to reflect a 1920 level of development would have reduced the average by about 1.3 million acre-feet per year to 16 million acre-feet per year.[18]

The assumption that all 17.3 million acre-feet were available with sufficient storage to save wet year flows for use in dry years had a second flaw that has become apparent in the decades since. Yes, large upstream reservoirs can catch excess runoff in wet years for use in dry years. But that necessarily results in large reservoir evaporation losses that would have to be deducted from water available at the Yuma gauge. The Fall-Davis report estimated that upstream reservoir evaporation would be slightly more than 2 million acre-feet per year.[19] The dams needed to provide the storage for the "conservation" Hoover called critical would further reduce the water available at Yuma to about 14 million acre-feet per year. Taking into account the growth in upstream depletions from 1902–20, upstream evaporation on major reservoirs and the 2.7 million acre-feet of existing uses below the gauge meant that the commission actually had only about 11.3 million acre-feet available for all new uses—just barely enough to cover the 10.8 million acre-feet of future needs as identified by the Reclamation Service but nowhere near enough to cover the 15 million acre-feet identified by the states.

At the end of the sixth meeting, New Mexico's commissioner Steven Davis offered a proposed compact based on an acreage limitation, and, to satisfy California commissioner William McClure, Hoover suggested it be made contingent on the building of a reservoir to store surpluses in wet years. Colorado's Carpenter and Utah's commissioner R. E. Caldwell could not be persuaded. The discussion continued during the seventh meeting, with Hoover offering a compromise based on irrigated acreage, but limited to twenty years. Again, led by opposition from Carpenter and Caldwell, the concept failed.[20]

The dissension and the inflated claims by the upper river states and Arizona disturbed Arthur Powell Davis.[21] It's likely that the upper river repre-

Table 2 Corrected water available at Yuma

1902–20 average flow at the Yuma gauge	17,300,000 acre-feet
Less uses in California and Mexico below gauge	2,700,000 acre-feet
Less correction for a common 1920 development	1,300,000 acre-feet
Less upstream reservoir evaporation	2,000,000 acre-feet
Actual water available at Yuma	11,300,000 acre-feet

Source: Authors' calculations.

sentatives knew their estimates were faulty and that the Reclamation Service estimates were more accurate, but politically they could not accept the federal numbers. Further, Carpenter had in mind an Upper Basin proposal for a compact that would avoid any numerical apportionments.[22] At that point, a proposal to apportion water based on the irrigation needs of the individual states was in serious trouble.

Neither Arthur Powell Davis nor the USGS's Nathan Grover made any attempt to educate the commissioners by correcting some of the flawed assumptions and help them reach a common technical understanding.[23] Perhaps since all seven commissioners and their advisors were comforted by Davis's view that there was sufficient water, and hoping for an agreement, they saw no reason to complicate the negotiations.[24] While perhaps providing a natural flow at Yuma may have been beyond their ability, the data and methodology to correct the average flows at Yuma or Laguna Dam to a 1920 level of development was available. LaRue had warned about this problem in Water Supply Paper 395. Further, in appendix B of the Fall-Davis report the flows at the Boulder Canyon Dam site were corrected, but like much that would have made a Colorado River agreement harder because it suggested there was less water in the river, it was ignored.[25]

With the failure of an acreage-based deal, Colorado's Carpenter, on behalf of the upper river states, made a counterproposal. Carpenter asked the lower river states to take whatever water arrived to the lower river after whatever future development that might occur upstream with no interference. He argued that the development and use of water was limited in the Upper Basin's mountains and canyons, and therefore the lower river was at minimal risk. He made a point that much of what was diverted in the upper river returned to the stream and was available for downstream uses. When pressed by Hoover and the lower river commissioners, Carpenter acknowledged that because transmountain diversions—water moved across the Continental Divide for use in eastern Colorado or on the Wasatch Front, outside the basin—were wholly consumptive, he would be willing to enter into a reasonable limitation on that use.[26] The lower river states rejected this proposal, noting that without some certainty as to how much water they would get, they would be unable to finance any projects.

The irony is that today, the Lower Basin actually might have been better off had they accepted Carpenter's offer with a limitation on exports out of the Upper Basin. The record suggests he would have accepted a limit several hundred thousand acre-feet per year less than what is now being exported out of the basin.[27] And he was right about the underlying issue. Due to topography and economics, the consumptive use from irrigation in the Upper Basin has been self-limiting.

The meetings that began in January in Washington ended in a stalemate.[28] The failure of the first three proposals for a compact left some commissioners frustrated. To New Mexico's Stephen Davis, it looked hopeless. "I feel that if we can't agree now, we can't agree at all, and it seems to me useless to have another meeting," he said.[29] After further discussion, in which Carpenter, Emerson, and Caldwell all encouraged their fellow commissioners to resume their deliberations at a future date, they returned home to hold a series of public meetings in their own states and consider how to break the stalemate.[30]

As expected, from the initial meeting the commissioners normally united as the upper river versus the lower river. This opened a path forward. During the commission's field hearing in Los Angeles, Arthur Powell Davis made the first formal proposal for a compact between basins, not states. His suggested two-basin approach would allow development to occur unrestricted in both basins for a period of fifty years. After this fifty-year period, any additional development in the Upper Basin would be subject to a priority call by existing projects in the Lower Basin. Another way to view the proposal was that all Upper Basin projects developed within this fifty-year window would have a senior priority to projects in the Lower Basin.[31]

At the next hearing, in Denver, the upper river states rejected this idea, and the field hearings ended without any progress to break the stalemate.

Led by Utah and Colorado, the upper river states remained insistent on a solution that would limit any restriction on their future development of Colorado River water.

Three things changed over the summer of 1922. The first was publication, in its final official form, of the Fall-Davis report. While draft material had circulated for two years, the final version of the report drew national attention to the serious flooding and water supply problems in the Imperial Valley and made a strong public policy argument for federal solutions. A large storage reservoir in Boulder Canyon or Black Canyon would provide flood control and water supply storage. Hydroelectric power, a much-needed resource for Southern California, could pay for the dam.[32]

Second, on June 5, 1922, the U.S. Supreme Court issued its ruling in *Wyoming v. Colorado*. In a unanimous decision, the court said that among prior appropriation states, the rule of priority applied on interstate streams. Colorado could not claim rights to a river's entire flow simply because the river started there. Applied to the Colorado River, this meant the faster-developing downstream states could lock up all the river's water. This was the result Colorado's Carpenter had feared. The decision was cheered in the Lower Basin and in Carpenter's words left the Upper Basin "badly exposed."[33]

The final factor contributing to the Upper Basin's change of heart was the introduction by members of Congress from California of legislation on April 25, 1922, authorizing construction of a federal dam at Boulder Canyon and the All-American Canal. The upper river states opposed the bill, noting the compact negotiations were not done. This gave them, in the short term, a powerful argument. But they understood that if the negotiations failed, publication of the Fall-Davis report would make it politically difficult to oppose future federal legislation aimed at solving the real and serious problems of the Imperial Valley.

Carpenter deserves most of the credit for recognizing that the Upper Basin would have to accept a legal limitation on its upstream development, then selling the idea to his fellow Upper Basin commissioners a concept that became the basis for the two-basin solution and would become the 1922 Colorado River Compact.[34]

After several months of delays the commission finally met again in Santa Fe, New Mexico, on November 9, 1922. Over the course of the next two weeks they hammered out the details of the compact. Time was short. Congress had given the commission only until the end of 1922 to finish.

While the talks continued between the end of the Washington meetings and the start of the Santa Fe meetings, there is no evidence the commissioners asked for any additional or clarifying hydrologic information from either Reclamation or the USGS about the Colorado River. The hydrologic misunderstandings and mistakes about how much water the Colorado River really had remained unquestioned, now locked into place.

At the commission's eleventh meeting, on November 11, Carpenter formally proposed the concept that would become the heart of the final deal—that the Colorado River Basin be divided into two subbasins with an equal division of the "mean or average annual established natural flow of said river . . . at Yuma, Arizona." Carpenter's proposed compact declared that flow to be 17.4 million acre-feet per year.[35] Considering the flow in the river's main stem and the contributions of its various tributaries, as summarized in the Fall-Davis report, Carpenter proposed that the Upper Basin be required to deliver an annual average of 6.264 million acre-feet at Lee Ferry.[36] Recognizing the variability of the river hydrology, Carpenter proposed the Upper Basin would deliver an average of 62,640,000 acre-feet every ten years.[37]

The confusion over how much water the river had and the failure to correct flows to a common level of development is on full display in Carpenter's proposal. It is clear he thought he was proposing an equal split of the river, but the proposal's hydrologic basis is flawed.[38] Was Carpenter proposing a fifty-fifty split of future consumptive water use on the river above the Yuma gauge, a fifty-fifty split of a reconstructed river under natural or virgin conditions, or was he simply putting a proposed deal on the table that had no clear underlying hydrologic basis? The language in his proposal seems to suggest that he thought he was dividing up the river based on natural flow.

However, the 17.4 million acre-feet per year average at Yuma was an average of actual gauged flows, not a natural flow. If his intent was to evenly divide up the water available for future growth, then it should have been corrected to reflect a 1920 level of development, about 16 million acre-feet per year. Further, since over 2 million acre-feet per year of California's existing uses were diverted below the Yuma gauge, his proposal of splitting 17.4 million acre-feet equally for each basin was beneficial to the Upper Basin. It would have provided 8.7 million acre-feet for new uses in the Upper Basin, but for the Lower Basin, the Imperial Valley's 2 million acre-feet of old uses would have had to have been supplied from the 8.7 million acre-feet, giving that basin only 6.7 million acre-feet for new uses.[39]

Despite the underlying uncertainty about what was meant, Carpenter's proposal was a breakthrough, setting forth basic principles that would end up in the final compact:

- The basin was to be divided into two subbasins with the dividing point near Lees Ferry;
- the apportionments of water for consumptive use would be to each sub-basin, not states;
- the Upper Basin would provide a ten-year minimum flow at Lee Ferry; and
- there would be an even split of any future obligation to Mexico between the two basins.

Overcoming initial opposition from the New Mexico and Arizona commissioners, they agreed to proceed with a compact based on two basins, not states. Although Arizona's Winfield Norviel agreed to the concept, he rejected Carpenter's numbers, saying 62.46 million acre-feet every ten years was too little for the Lower Basin. After consulting with both basins, Hoover suggested a counteroffer—82 million acre-feet every ten years. Hoover's proposal was based on the assumption that the actual flow of the river at Lee Ferry from 1899–1920 was 16.4 million acre-feet per year. Since there was no gauge data for Lee Ferry at the time, the 16.4 million acre-feet number was based on the Laguna Dam data from the Fall-Davis report, and a simplified assumption that tributary gains to the Colorado River below Lee Ferry and evaporation and system losses between Lee Ferry and Laguna Dam were the about same. Thus, flows of the river at Laguna Dam and at Lee Ferry were approximately the same.[40]

Hoover's counterproposal was different than Carpenter's for several reasons: First, Hoover's proposal was based on a fifty-fifty split of the river at Lee Ferry, not a fifty-fifty split of the river at Yuma.[41] Second, by suggesting a fifty-fifty split of the water available at Lee Ferry, the proposal left the

Gila River out of the calculation. In a nod to Norviel, Hoover reckoned that this would leave the full use of the Gila River system to Arizona—a shift in approach to the river's allocation that would echo through the decades that followed.[42]

Again, there were problems with Hoover's hydrologic assumptions. The Laguna Dam numbers were not corrected to a 1920 level of development. Had the commissioners been given such a number, the average annual flow would have been at least 800,000 acre-feet less.[43]

The Upper Basin commissioners rejected Hoover's 82 million acre-feet, offering 65 million acre-feet every ten years. With the bookends on the delivery at Lee Ferry set, the bartering continued. The commissioners ultimately agreed to a requirement that the states above Lee Ferry not deplete the flow at Lee Ferry below a ten-year moving total of 75 million acre-feet.[44] This would become Article III(d) of the compact. It's worth noting that 75 million is about halfway between 65 and 82, and it's close to what Hoover might have proposed had he been using development corrected average flows at Laguna Dam.[45]

Today the small differences between the numbers loom large, impacting river operations at a narrow margin that means the difference between a sustainable supply and dwindling reservoirs. But the next piece of the negotiations offers important but little-remembered historical context—the commissioners decided to break from the Carpenter proposal in an important way. Carpenter's proposal was to split the entire river. Instead, they decided to split the river in three ways: an apportionment for each basin and a third portion that would be set aside for future apportionment. In a crucial piece of the story that is largely ignored today, the commissioners were assuming that they still had a substantial share of water left over, to be divided later among the states.

Hoover, speaking during the eighteenth meeting on November 16, explained it this way: "Yesterday we got away from the point of view of a fifty-fifty division of water. We set up an entirely new hypothesis. That was that we make, in effect, a preliminary division pending the revision of this compact. The seven and a half million annual flow of rights are credited to the South, and seven and a half million will be credited to the North, and at some future day a revision of the distribution *of the remaining water* will be made or determined" (emphasis added).[46] This was a major shift in the structure of the compact. Under this proposal, the 7.5 million acre-feet per year of beneficial consumptive use provided each basin would include all existing uses. Later in the negotiations, to address Norviel's concerns about protecting Arizona's use of the Gila, the Lower Basin's apportionment was expanded by a million acre-feet per year to a total of 8.5 million acre-feet. Most importantly, this course of action created a third category of water, a surplus that would be divided in the future. This move gave them some breathing room. If the initial apportionments were insufficient, the compact would provide for a process to fix them in the future.

Based on the commonly held understanding of the commissioners that the river would support 20–22 million acre-feet of development, it meant the expectation of a surplus available for future apportionments and for Mexico was 4–6 million acre-feet. The existence and use of this surplus would be a major factor in many of the future decisions and fights that would shape the river.

Based on the hydrology presented in the Fall-Davis report, the assumption of a future surplus was understandable. They all believed they had a 20-to-22-million-acre-foot river. By avoiding LaRue's concerns about the sufficiency of the water supply—ignoring reservoir evaporation, failing to correct their flow data for development that had already happened on the river, and other sins—they had conjured up a larger Colorado River than nature could actually provide. There would be, they believed, a surplus.

Once the commissioners had agreed on the Upper Basin's flow obligation at Lee Ferry, the concept of initially apportioning only a portion of the river's available supply, and including existing uses in the initial apportionments, a number of difficult and contentious hydrology issues remained. First, Arizona's Norviel remained convinced that the initial proposal to apportion 7.5 million acre-feet per year to each basin was insufficient to cover both Arizona and California's needs from the main stem and allow Arizona to fully use of the Gila River. Second was the issue of how to deal with a future treaty with Mexico. Third was how to address the rights of the region's many Native American communities and reservations. The final issue was question of how the compact would address storage.

The Gila River arises in the mountains of southwestern New Mexico, then flows into Arizona. Its biggest tributary, the Salt River, drains the rim country of Arizona. The two meet just west of Phoenix. From the Phoenix area to Yuma the river flows through one of the hottest and driest deserts in North America. In its natural state, the Gila and its tributaries produce, on average, over two million acre-feet per year as they leave the high country, but by the time the river arrives at Yuma half its waters have been lost to native vegetation and the hot desert air.

Modern large-scale irrigation in central Arizona began in the 1890s. The Reclamation Service's Salt River Project in the early 1900s accelerated the development of what would become a major farming region. When Winfield Norviel arrived in Washington, D.C., in early 1922 for the first commission meeting, over 430,000 acres of land were under irrigation.[47] Mining and agriculture were the backbones of the state's economy. The question was whether that Gila water would be counted against Arizona's compact share. By mid-November, as the commissioners were meeting in Santa Fe, they all understood that if the Arizona legislature was going to ratify the compact, Norviel needed protection for the Gila.

The technical roots of the solution are murky, but there are clues. The Fall-Davis report put the average flow of the Gila River at its mouth as 1.07 million acre-feet per year, based on measured flows at the Dome gauge for 1903–20.[48] Like all other reported flows in the Fall-Davis report, it was not adjusted to a common 1920 level of development nor was it a natural flow. Further, because of the rapid development on the Gila from 1900 to 1920, the big natural losses compared with its natural flow, and some exceptionally large flow years, the 1.07 million acre-foot figure was misleading. It had little relevance to the Gila River as of 1920.[49] And yet, like other numbers in the Fall-Davis report, whether technically justifiable or not, the number provided a foundation for a deal.

It was Nevada's James Scrugham who proposed increasing the Lower Basin's apportionment by an additional million acre-feet per year as a way to address Norviel's concerns.[50] The Fall-Davis report's Gila number gave the commissioners a way to satisfy Arizona's concerns without opening a hornet's nest regarding how much water the Gila really had to offer or how much of it Arizona was actually using. Simply increasing the Lower Basin's apportionment by a million acre-feet allowed a gentleman's agreement that they did it to satisfy Arizona.[51] An additional million acre-feet per year was big enough to give the Lower Basin a cushion and small enough to not alarm the Upper Basin.

Norviel took comfort in the combined effects of Article III(d), which obligated the States of the Upper Division to not deplete the flow of the Colorado River below 75 million acre-feet over any ten consecutive years, with Scrugham's additional million acre-feet per year. Norviel reasoned that the Upper Basin's average delivery of 7.5 million acre-feet per year would supply the Lower Basin's III(a) 7.5-million-acre-foot apportionment, leaving the additional million available for further development of the Gila River.[52] This additional 1-million-acre-foot apportionment would become Article III(b) of the compact.

The Gila River problem and the 1-million-acre-foot compromise raised a fundamental question that bedevils river management to this day: How are the compact apportionments to be measured? Different approaches to the seemingly simple question yielded wildly different results—one approach favoring Arizona's use of the Gila, the other favoring California's allocation at Arizona's expense.

While a future treaty with Mexico was politically sensitive, most of the commissioners and their legal advisors believed it was a real possibility. They then had to provide some foundation in the compact for apportioning obligations to Mexico. Everyone understood that Mexico's future water requirements would not be small. The Reclamation Service estimated that Mexico's future additional acres for irrigation were over 600,000, and both Reclamation and the states believed its total water requirements could exceed 3 million acre-feet per year.[53] The decision of the commission to leave a surplus for future apportionment gave them a way to address the issue. Any water provided to Mexico under a future treaty would first come from the surplus, estimated in 1922 to be about 4–6 million acre-feet per year. If the surplus was insufficient, then the deficiency would be shared equally between the two basins, and the States of the Upper Division would deliver the Upper Basin's share on top of its obligation not to deplete the flow at Lee Ferry below 75 million acre-feet every consecutive ten years. There is still no consensus about how to interpret the article, but to date it has escaped litigation.

The commissioners addressed existing and future rights of Native Americans in Article VII with the simple statement that "nothing in this compact shall be construed as affecting the obligations of the United States of America to Indian Tribes." The reality that the United States was a trustee for the rights of tribes and that tribes potentially had rights under the U.S. Supreme Court's 1908 decision in the case of *Winters v. United States* was well understood at the time of the compact's negotiation. Although Hoover would later testify that this provision may have been unnecessary, he considered it one of the core federal interests.[54] Hundley notes that no effort was made to determine how many tribes were in the basin and what their water rights might be. Further, the commissioners believed that, in any event, such rights would be "negligible."[55]

While Norviel's critical issue was the Gila River, for California's McClure it was the construction of storage to reregulate Lower Basin flows and control the devastating floods that imperiled the Imperial Valley. The construction of a storage reservoir, for flood control and other purposes, on the lower Colorado River below the mouth of the Grand Canyon was a major political objective for the State of California. The need for such a reservoir was not disputed by the Upper Basin commissioners. However, these commissioners, especially Carpenter, were opposed to including a provision requiring storage in the compact. Carpenter viewed the compact as a legal document, defining rights and obligations, not a policy or implementation document. This left many Californians, especially those from the Imperial Valley and the City of Los Angeles, bitter.[56]

However, without a storage provision, the compact had a technical problem. Under III(d), the States of the Upper Division are only obligated to deliver 75 million acre-feet every ten years. During the debate over the ten-year flow obligation at Lee Ferry, Arizona's Norviel sought an annual minimum flow, but the other states' commissioners would never agree to one.[57] This meant that in dry years, without storage, the lower river could be dried up, depriving some of the most senior rights of their accustomed water supply. In theory, the Upper Basin could deliver no water for a period of time without a compact violation. It would just have to deliver more at some future time during the ten-year window.[58]

To satisfy both Carpenter and improve the chances California would ratify the compact, the commission crafted an innovative compromise from a suggestion made by McClure to protect existing rights.[59] He proposed a provision that present perfected rights would be "unimpaired" by the compact, but once 5 million acre-feet of storage was constructed for the benefit of the Lower Basin, then any claims that Lower Basin rights might have against water rights in the Upper Basin would be satisfied by this 5 million acre-feet of storage.

Under this approach, the compact would not require storage or any other structure, but if 5 million acre-feet of storage was not built, then existing rights on the lower river that might have a priority claim against users in the Upper Basin before the compact was signed, like the Imperial Irrigation District's right for 10,000 cfs, since they were unimpaired by the compact, were still legally free to do so.[60] Hoover and McClure reasoned that this provision would give the Upper Basin states strong incentive to support the authorization of a dam. Although this initially failed to satisfy most Californians, the commission was right.[61] The Upper Basin states actively supported the Boulder Canyon Project.

The commission approved the Colorado River Compact at their twenty-seventh meeting at Bishop's Lodge on November 24, 1922. The commissioners signed the compact later that evening in a ceremony at the Hall of Governors in Santa Fe.

The commissioners began the process of the negotiations with an incomplete understanding of the hydrology of the Colorado River. Their knowledge and understanding improved through the course of the negotiations, but learning and understanding "the facts" or the science of the river was never really more than interpreting information provided by the Reclamation Service and the USGS in a way that met their negotiating needs.[62] Their basic understanding on the day the commissioners signed the compact was the same as what they heard from Arthur P. Davis on the first day: with conservation (meaning storing water for future use) there was sufficient Colorado River water to meet all anticipated future needs.

What the commissioners believed on November 24, 1922, was that there was enough water to reach a fragile political agreement that, once ratified, would clear the path for development. What they didn't understand, but could have had they wanted to, was that 1900–1920 was a period of extraordinary wetness for the Colorado River. However, in the next several years and before the compact would be ratified, a few inquisitive USGS hydrologists, an engineering professor, and a review board of engineers and geologists all concluded that the basic assumptions concerning the flow of the river used by the commission were faulty and too high.

They would be ignored.

CHAPTER 4

Selling the Compact

In a meeting of the Wyoming state bar association in January 1923, an argument broke out over the ratification of the Colorado River Compact. The question: How much water did the Colorado River really have?

Wyoming has always played a relatively small role in the hydrology and history of the river. Its neighbor Colorado dominated the upper reaches of the Colorado, both in the water it supplied and used and in the politics. But there in Cheyenne, far outside the Colorado River Basin, one of the most important arguments about the future of the river was playing out.

Development of the Colorado was entering its next phase, as the small group that negotiated the Colorado River Compact dispersed to sell the deal. They faced a fresh round of questions about what the deal would mean to the aspirations of water users in the Colorado River Basin. How did they come up with the numbers that had solidified in the compact's allocations? What would the ultimate obligation of the Upper Basin be to deliver water downstream to the lower deserts? What of the future water needs of a possible treaty with Mexico? How might the water already used in the basin affect the numbers used to cut the deal? And what of the surplus—the expectation codified in Article III(f) of the compact that after all the West's dreamed-of farms and cities were built, along with the infrastructure to deliver their water, they would reconvene in half a century to divide up the rest of the water?

The months that followed saw the most detailed public discussions to date of the relationship between the Colorado River's hydrology and the number and size of the future farms and cities it might support. But the discussions did little to break through the screen of optimism thrown up by those who favored massive water projects and the expansion of the basin's agricultural lands. Time and again, compact backers threw up optimistic scenarios—plenty of water for each state's dreams, with a surplus for the future.

Only in Wyoming was the otherwise unasked question raised, the one first broached by E. C. LaRue in Water Supply Paper 395, his *Colorado River and Its Utilization*: Was there really enough water to do all this? The question came from a lawyer named Nellis E. Corthell. An attorney who had represented Wyoming in its legal battle with Colorado over the Laramie, Corthell was skilled in water law but modest about his knowledge of the Colorado. "My personal acquaintance with the River," he wrote, "is very slight."[1] But his critique was prescient. It appeared, Corthell told a gathering of lawyers and state legislators at that Cheyenne meeting, that the Colorado River did not have enough water to meet the allocations set out in the compact. Based on an insightful analysis of the Fall-Davis data on which the deal was based, Corthell questioned the compact allocation and called for a conservative approach. With imperfect data and the risk that the short time period being used was not representative of the river's long-term flow, Corthell urged caution. "It is of the very highest importance to exercise care and circumspection in apportioning the water between the two divisions," he would write that summer, "since any mistakes in this allocation would be irreparable and indeed may not become apparent for many years."[2]

Returning Home

After the compact signing, Delph Carpenter, Herbert Hoover, and Arthur Powell Davis returned home proud of what they had accomplished. In future years, Hoover and Carpenter would each claim credit for being the mastermind of the compact, but much of the credit should go to Davis. During the negotiations, he had been Hoover's right-hand man. After his initial approach to move the states toward a compact based on irrigated acreage failed, he was the first to suggest the two-basin approach. He kept the commissioners focused and gave them no reason to doubt there was sufficient water to meet their needs. He had set in place the political foundations that would both launch the large-scale development of the Colorado River's water and put his agency in the position to become, in a few decades, an engineering giant.

While in Washington, D.C., Davis was called on to answer hydrology and engineering questions concerning the pact, but it didn't take him long to plot his return to the Colorado River Basin. He intended to join E. C. LaRue and Herman Stabler on the 1923 Birdseye river expedition that would make detailed investigations of Colorado River dam sites. He never made the trip. Instead Interior secretary Hubert Work, a medical doctor turned master bureaucrat,[3] decided that the tiny Reclamation Service, designed and managed by Davis to help reclaim arid lands for small farmers, needed to go big. To build the Boulder Canyon Project—what would become Hoover Dam—and future giant projects like the Grand Coulee and Shasta Dams, Work needed a major bureaucracy and a professional manager. On June 19, 1923, Work replaced Arthur Powell Davis with David W. Davis (no relation).[4] In 1924, David Davis was replaced by Elwood Mead. Under Mead's tutelage the Bureau of Reclamation would soon build both the world's tallest dam (Hoover Dam) and its largest hydroelectric power plant (Grand Coulee).

Ratification by Six States

In the months that followed the compact signing, Hoover and the state commissioners began grappling with the implications of what they had done. In reports written by the commissioners and their technical advisors, they tried to answer questions about how much water the Colorado River had, how much future water users might want or need, and how the resulting allocation rules would affect each state. But their efforts were more about winning political support for the deal than about shedding honest light on the Colorado River's hydrology.

These testimonials and reports nevertheless provide valuable insights into the hydrologic assumptions and understandings the commissioners and their advisors had. They confirm that each believed that the Colorado River had sufficient water to accommodate at least 20 million acre-feet of development. During the six-year struggle for ratification, hydrologic studies would surface, including an expansion by LaRue of his analysis of the droughts of the nineteenth century, that would call into question the basic assumption that the river's water supply was sufficient to meet the intent of the compact. But the development community remained unmoved.

The commissioners from Colorado, Wyoming, and Utah submitted detailed reports and recommended prompt ratification. Commissioners from New Mexico, Nevada, and California submitted shorter reports, also recommending prompt ratification. Arizona's Norviel, recognizing that Arizona's political winds had shifted with the election of Governor George W. P. Hunt, didn't officially submit a report, but published his analysis of the compact and urged ratification in a mining journal.[5]

Hoover, still serving as secretary of commerce as well as being the Colorado River Compact Commission's chairman, spent considerable effort pursuing congressional ratification and helping state commissioners obtain ratifications from their legislatures. He both submitted a report to Congress and responded to detailed questions from Rep. Carl Hayden of Arizona. The responses to Hayden, which were given careful consideration and scrutiny, are one of the most important accounts of the federal government's views at the dawn of the development of the Colorado River.[6]

Prior to the 1920s, the federal government had played only a minor role in the nation's water management, a task that fell almost entirely on state governments and happened at more local scales. That made questions about the new, expanded role for the U.S. government contemplated by the 1922 compact and the big dams and canals that would follow central to the ratification process. In his report to Congress, Hoover focused on the federal role, and, true to the time, he viewed the federal interests narrowly. Federal interests, he explained, were limited to navigation, international relations (with Mexico), as proprietor of public lands and owner of irrigation works, as trustee for Indian tribes, and in power plant regulation and permitting under the Federal Power Act. Except for navigation, where he made the case that the Colorado River was no longer navigable, he told Congress that the compact protected all of the federal interests.[7] Over the next decade, and by the time his term as president ended, the federal role would undergo a seismic shift. Under the Bureau of Reclamation and Army Corps of Engineers, the national government would become the promoter, builder, and operator of major projects. Reclamation also, as we shall see, came to dominate the development of the science used to support its mission.

Arizona's Hayden was more interested in the specifics of the compact and how it affected his state. He asked whether the apportionments were arbitrary or based on the actual needs of each basin. Hoover made clear he believed they were not arbitrary and that they were based on a "careful consideration of the respective needs"—existing projects and also those "considered economically feasible and also those of doubtful feasibility" during the next seventy-five years.[8]

In one of the most important interactions on the record, Hayden asked Hoover to explain the apparent mathematical relationship between the 7.5 million acre-feet apportionment to the Lower Basin under III(a) and the Upper Basin's 75-million-acre-foot obligation at Lee Ferry under III(d). Hayden was referring to the apparent numerical relationship between the 7.5 million acre-feet per year and 75 million acre-feet over ten years.

In his response, Hoover tries to disconnect the two, noting that the 7.5-million-acre-foot apportionments were based on an assessment of needs, whereas the III(d) obligation to deliver water at Lee Ferry was a "definite quantity of water which must pass that point." He further states that "in the improbable event of a deficiency, the Lower Basin has the first call on the water up to a total of 75,000,000 acre-feet each 10 years. . . . The period of 10 years was fixed as a basis of measurement, as being long enough to allow equalization between years of high and low flow." The commission minutes support Hoover's interpretation that it was more numerical coincidence than linked plan. The 75-million-acre-foot requirement was negotiated when the commissioners were focused on splitting the entire river between the two basins. Agreement on the 7.5-million-acre-foot apportionments in paragraph III(a) occurred later in the deliberations and after they had decided to apportion only a portion of the river's supply. The 7.5-million-acre-foot numbers were based on projected demands from the Fall-Davis report with a reasonable cushion.

Hoover's comment that the Lower Basin has the first call on the water up to 75 million acre-feet has significant implications for the Upper Basin. Since the 1922 compact was ratified, water officials in the Upper Basin have occasionally suggested that if there is not enough water for the Upper Basin to both develop 7.5 million acre-feet per year while still allowing 75 million acre-feet to pass Lee Ferry every ten years under Article III(d), then III(a) would control, and the Upper Basin states could legally deplete the flow to less than 75 million acre-feet every ten years. By making clear the compact framers' intentions, Hoover's answer puts a dagger in this theory.[9] To give the dagger a bit of a twist, the evidence is that the Upper Basin's commissioners either agreed with or saw no reason to disagree with Hoover. Hoover's testimony was made available to all of the commissioners, and there is no record it was ever challenged. The Upper Basin's silence on this question suggests its representatives genuinely believed the high-end estimates of the river's flow, on which the deal had been based, and saw no risk in a compact that obligated them to deliver 75 million acre-feet of water every ten years. The comments of both Hoover and Davis were printed in the *Congressional Record*, and Colorado's Carpenter refers to them in his supplemental report.[10]

Hayden asked about the implications of a future treaty with Mexico asking if "paragraph (c) of Article III contemplate(s) a treaty between the United States and the Republic of Mexico under which one-half of a deficiency of water for the irrigation of lands in Mexico shall be supplied from reservoirs in Arizona?" Hoover responded in the negative. "No. Paragraph (c) of Article III does not contemplate any treaty. It recognizes the possibility that a treaty may, at some time, be made and that under it, Mexico may become entitled to the use of some water, and divides the burden in such an event." In response to a question about the obligations of each of the basins to Mexico, Hoover answers that "it is provided in the compact that the upper states shall add their share of any Mexican burden to the delivery to be made at Lee Ferry." He makes it clear that the obligation of the Upper Basin to help meet a possible Mexican treaty obligation under Article III(c) is additive to Article III(d) regarding upstream states' obligations at Lee Ferry.[11]

Finally, Hoover made the most important promise of the river's hydrology—
that there was not only enough water for all the projects contemplated at the
time, but that such development would leave unappropriated water for the
taking for future generations for the future. "The unappropriated surplus,"
Hoover said, "is estimated at from 4,000,000 to 6,000,000 acre-feet, but
may be taken as approximately 5,000,000 acre-feet."[12] He clearly believed
that there was enough water for all, something echoed in many of the state
reports written in support of ratification.

Hayden also had questions for Reclamation Service Director Arthur Powell Davis that began to reflect an expanded idea of how to measure and think
about the Colorado River's flow. For the first time Davis, without explicitly
acknowledging doing so, introduced the idea of calculating the river's "natural flow"—the amount of water that would have flowed absent upstream
diversions for human use.[13] The importance of calculating "natural flow" is
one of the foundations of modern Colorado River management, but in the
1920s it was a new and poorly understood concept. His answer is the first
documented reference to a reconstructed natural flow of the Colorado River
at Lee Ferry. No tables of natural flows are included in the Fall-Davis report,
and while this now commonly used metric was discussed in concept by the
commissioners, it was not widely used during the negotiations.

Davis did it in response to Hayden's question about the river's hydrology:
"After deducting the maximum quantity of water that may be diverted out
of the upper basin and the maximum amount that may be consumed by
irrigation and domestic uses, what is your estimate of the average annual
run-off from the upper basin in acre-feet at Lee Ferry?"

Davis summarized it in a table (see table 3). Arthur Powell Davis's assumption that the mean discharge at Lee Ferry and Laguna were about the same
was common at the time. Based on what we know today, it is a reasonable

Table 3 Arthur Powell Davis's natural flow calculation

	Acre-feet
Mean discharge at Lee Ferry 1903–20 (assumed the same as Laguna)	16,400,000
Past depletion, Upper Basin, 1,094,000 irrigated acres	1,700,000
Total: Reconstructed at Lee Ferry	18,100,000

approximation. Further, unlike the Fall-Davis report and compact negotiations, Davis fully recognized that depletions from 1899 to 1920 were steadily increasing; thus, he corrected the flow at Laguna for the average Upper Basin depletion over that period.[14]

Davis assumed that the maximum consumptive use in the Upper Basin would only be 6.59 million acre-feet per year, almost a million acre-feet short of the 7.5 million acre-feet per year apportioned to the Upper Basin under Article III(c). The 6.59-million-acre-foot figure for the build-out of the Upper Basin would be used again in hydrologic studies, several of them vital to the ultimate development of the river. This number did not surprise or upset anyone in the Upper Basin. They believed that the 7.5-million-acre-foot apportionment in III(a) was based on their anticipated uses plus a reasonable cushion.

Based on what we see today, Davis' estimates of the diversions out of the basin and the in-basin consumptive use attributable to irrigation were both off by a factor of two, but not in the same direction. His estimate of irrigation use of 6.15 million acre-feet per year is over twice the 2.7–2.8 million acre-feet now being consumed for irrigation in the Upper Basin. His estimate of 444,000 acre-feet per year for exports is only about 50 percent of today's actual exports of 750,000 to 900,000 acre-feet.[15]

Only three weeks after Carpenter signed the 1922 compact, he submitted a detailed report recommending ratification to Colorado governor Oliver Shoup. In preparing the report, Carpenter was assisted by Ralph Meeker, Colorado's deputy state engineer and his principle engineering advisor during the negotiations.[16]

Carpenter was given great deference by legislators not only in Colorado but throughout much of the basin. He argued that the total water available in the Colorado River, including existing uses, was 20.5 million acre-feet per year—17.5 million acre-feet originating above Lee Ferry, with the remaining 3 million acre-feet below Lee Ferry. Since the compact apportioned 16 million acre-feet, Carpenter argued, there was a 4.5-million-acre-foot surplus. He noted that the Upper Basin was consuming 2.5 million acre-feet, leaving 5 million acre-feet per year for future development.[17]

Carpenter's estimated natural flow at Lee Ferry of 17.5 million acre-feet per year is 600,000 acre-feet per year less than the 18.1 million acre-feet used by Arthur Powell Davis, but still almost 3 million acre-feet per year more than today's understanding that the 1906–2016 average was 14.8 million acre-feet per year. His estimated surplus of 4.5 million acre-feet is slightly less than Hoover's estimate of 5 million acre-feet, but still large enough to make his case that there was ample water left over for both Mexico and future apportionments under Article III(f), the compact provision that set aside a surplus for later allocation. In a letter to Wyoming's Emerson, Arthur Powell Davis agreed with Carpenter's estimated total water supply of 20.5 million acre-feet.[18]

Carpenter addressed the flow requirement at Lee Ferry by concluding that "it is evident that the States of the upper basin may safely guarantee 75,000,000 acre-feet aggregate delivery at Lees Ferry during each 10-year period." Here he refers to the obligation as a guarantee.[19]

Like Carpenter, Wyoming's Frank Emerson and Utah's R. E. Caldwell also made detailed reports to governors and legislatures.[20] Both Caldwell and Emerson held the title of state engineer from their respective states and thus were experts on water supply and river hydrology. Their reports were similar, suggesting no doubt about the water supply estimates on which the compact had been based. It showed up most clearly in their analysis of the compact's Article III(d) provision, which set out the standard of 75 million acre-feet flowing past Lee Ferry every ten years. If ever there were a doubt about the hydrology, this would have been the place it would have been most likely to show up, because a shortfall in the river's hydrology would place Upper Basin compliance with III(d) at risk. "It will be impossible under any conceivable circumstance for the Upper States to prevent 75,000,000 acre-feet going past Lee's Ferry in any ten-year period," Caldwell wrote. Even with wildly optimistic estimates for the future expansion of irrigated agriculture, Caldwell believed there would be as much as 6 million acre-feet of surplus to divide later.[21] Emerson expressly ties Article III(b), the extra 1 million acre-feet of water apportioned to the Lower Basin, to Arizona's future use of the Gila River.

Nevada's Scrugham, New Mexico's Steven B. Davis and California's McClure submitted short reports to their states, all recommending ratification. Arizona's Norviel never submitted a formal report to his governor or legislature. Upon his return to Arizona he faced a newly elected governor, George W. P. Hunt, who opposed the compact. Norviel and his legal advisor, Richard E. Sloan, published statements concerning the compact in the *Arizona Mining Journal* on January 15, 1923.[22]

Both Norviel and Sloan discuss the hydrology of the river using the commonly held assumption that the river system would provide over 20 million acre-feet of water for development and use. Using the Yuma gauge data, Norviel concludes, "Adding the amount disclosed at Yuma to the amount diverted and used above, including that from streams in Arizona, would make production of the basin between 22,000,000 and 24,000,000 acre-feet per annum."

All of the basin state legislatures except Arizona's, with only a few controversies, ratified the compact by the end of their 1923 sessions. The only serious questions about the hydrology came from Corthell in Wyoming. He hinted at what would later become clear as the biggest shortcoming of the hydrologic analysis on which the compact was based—the failure to come to terms with deep droughts that had happened in the past. "The known imperfections of measurement and observation, as well as the enormous variation in stream flow which is known to occur over long periods of time suggest caution," he told the bar association audience in the January 1923 Cheyenne meeting. He also criticized the compact's treatment of a future obligation to Mexico. He believed that any water provided by treaty to Mexico should be deducted from the Lower Basin's apportionment. His most prophetic opposition was that he believed the compact negotiators overstated the amount of water available for development. He thought the amount available from the entire river was between 17.5 and 18.5 million acre-feet, not the more accepted 20–22 million acre-feet.[23] Corthell based this criticism on the 1920 water year, which had a reconstructed flow of only 16 million acre-feet and his view that the "twenty year average was not safe and reliable."[24] Wyoming state engineer Emerson defended the numbers, and enlisted Arthur Powell Davis in his defense. The Wyoming legislature deferred to Davis's expertise over Corthell's argument and overwhelmingly ratified the compact.

In Colorado, some questions were raised about the interpretations of Articles III(b) (the extra million acre-feet to the Lower Basin) and VIII (the provision intended to encourage early development of a reservoir), but with the help of his fellow commissioners Carpenter easily addressed them. Of more interest, a state legislator from Southwestern Colorado used E. C. LaRue as a source and questioned Carpenter's water supply assumptions.[25] This marked one of the only places in the entire compact discussion to date that the questions raised by the work of E. C. LaRue, the nation's most prominent Colorado River hydrologist, came up. Carpenter waved the concerns away, pointing to the conclusions of his advisor R. I. Meeker and implying that Meeker knew more about the Colorado River hydrology than LaRue. The Colorado legislature then unanimously ratified the compact.[26]

In California, bitterness remained over the lack of compact language requiring construction of a major reservoir on the lower river. Its legislature debated whether or not its ratification should be made contingent upon federal legislation authorizing the construction of storage at Boulder Canyon and an All-American Canal. Ultimately, with support by the Imperial Irrigation District board of directors for a clean ratification, the California legislature followed Colorado and unanimously approved the compact, without reservations.[27]

Arizona's opposition was led by two strong-willed individuals: newly elected Governor Hunt and irrigation promoter George Maxwell. Their opposition was based on a number of factors, including state sovereignty, loss of future power revenues, and concerns and confusion over how the compact dealt with the Gila River. Maxwell believed the compact would rob Arizona of sufficient water for his scheme to divert enough water from the river on the western end of the Grand Canyon to central Arizona to irrigate several million acres of desert land. Hunt also wanted to await the results of an engineering study of irrigable lands in Arizona. On July 5, 1923, the report was completed. The engineering team was led by USGS hydraulic engineer E. C. LaRue, author of Water Supply Paper 395. The report concluded that a project to irrigate two million acres of land in Arizona with a several-hundred-mile supply canal from the western Grand Canyon was possible, but economically impractical.[28]

While the report cast major doubts on the economics of Maxwell's "High Line Canal" project, it hardened Governor Hunt's opposition to the compact. Hunt focused on an addendum to the report authored by LaRue that suggested that although the project was not economical at the time, in the future, the scheme might be feasible. LaRue also concluded the Arizona project, along with other potential Lower Basin irrigation projects, could require a water supply of 16 million acre-feet of water, almost the entire yield of the Colorado River. LaRue implied that there was no real surplus under the proposed compact and that the river was, in fact, overallocated.[29] Carpenter was critical of LaRue and his impact on Arizona's attitude toward the compact. In an undated memo titled "E. C. LaRue," he refers to LaRue's influence on Arizona's ratification of the compact as "very destructive."[30]

The dispute became personal. Hoover considered Maxwell a "crackpot" and had little respect for the populist Hunt. One Utah congressman went so far as to say Maxwell "sounded as if he had been chewing peyote."[31] There was support in Arizona for ratification as well, including Representative (and soon-to-be Senator) Carl Hayden and the Arizona Republic. A vote in the Arizona House for a clean ratification lost on a 22–22 tie vote. After this close vote in 1923, Arizona's opposition hardened. Hunt was reelected in 1924, and California's plans for additional large diversions from the Colorado River to the Los Angeles area would become Arizona's major concern.

Initially, the other states were patient, giving supporters of the compact within Arizona time to make their case. This did not work. Hunt's reelection in 1924 led Carpenter to the conclusion that the other six states needed a six-state ratification strategy.[32] The vehicle to accomplish a six-state approval was the federal legislation needed to authorize construction of the large Colorado River dam that was central to the project, and an "All-American" canal route from the river to the Imperial Valley. California congressman Phillip Swing, who represented the Imperial Valley, and California senator Hiram Johnson were primary sponsors of the "Swing-Johnson bills." It took years, and multiple tries. The pair introduced the first in 1922, during the negotiations. The second and third came in 1924, before they introduced the fourth and finally successful version in 1927, during the ratification process. Arizona's congressional opposition was fierce, becoming a test of loyalty for Arizona politicians.[33]

The six-state ratification worked well for Swing and Johnson. It gave them an opportunity to tie the authorization of the Boulder Canyon Project and the All-American Canal to congressional ratification of the compact. The political stage was now set for six of the seven basin states to support a Boulder Canyon Project Act.

CHAPTER 5

The Troubling Science
of E. C. LaRue

Sitting before a panel of U.S. senators in December 1925, hydrologist Eugene Clyde LaRue was blunt. The senators were on the brink of launching the largest water development scheme the nation had ever seen, to build the first of a generation of dams and canals in the Colorado River Basin that would, in the words of boosters, turn "four million acres of desert into a pleasant countryside."

It was to be one of the great projects of a growing nation, bringing modern science and engineering to bear in service of progress. "The harnessing up of this river to do the work of man," the *New York Times* gushed, "is destined to be one of the greatest engineering undertakings since the pyramids were reared in Egypt's sand. . . . The backlands in seven of our largest and most undeveloped States will be open to homebuilders, an empire greater than France."[1]

Yet here was LaRue—the most accomplished of the nation's Colorado River experts, the epitome of the sort of technical expertise on which the great enterprise was to be founded—suggesting the science of the Colorado River told a very different story.

"For many years," LaRue told the senators, "it has been reported that there was plenty of water for all." LaRue then turned to a report he had written nine years earlier. "Evidently," he read from his 1916 report, "the flow of

Figure 2 Eugene Clyde LaRue measuring the flow in Nankoweap Creek, 1923. Courtesy of USGS.

Colorado River and its tributaries is not sufficient to irrigate all the irrigable lands lying within the basin."[2] Since that report, new work had made the risk even more clear, LaRue said. A repeat of a drought of the magnitude seen from 1886 to 1905 would create annual water shortages in the millions of acre-feet per year. The dream of turning those desert acres into "a pleasant countryside," LaRue was telling the senators, would inevitably collide with hydrologic reality.[3]

What happened to LaRue and his findings about the Colorado River Basin's water supply is a case study in the role of inconvenient science in political debates. LaRue and his work had been hovering in the background of Colorado River political discussions for more than a decade. After the 1916 publication of Water Supply Paper 395, he continued his studies of the Colorado River. By 1920, his view that the water supply available from the river was insufficient to meet the future needs was well known in the basin and by his Interior supervisors. He focused his energies on two areas. The first was advancing the understanding of the hydrology of the river by reconstructing flows back to 1851, allowing consideration of not only the wet decades of the early twentieth century, on which the compact's negotiators relied in allocating the river's water, but also the dry decades that went before. The second was a comprehensive development plan for the Colorado River to maximize supplies by minimizing reservoir evaporation.

LaRue was not alone. In 1924, his USGS colleague Herman Stabler authored a report concluding that the river's annual flows from 1878 to 1900 were far below those seen from 1900 to 1920. Stabler's work was endorsed by a well-recognized irrigation engineer, George E. P. Smith from the University of Arizona.

This information put the compact commissioners in a tough spot. There was now credible science that the river's long-term flows might be much lower than what they assumed. Yet in the short term, conditions on the Colorado River remained wet. Pushed by U.S. commerce secretary Herbert Hoover and Colorado lawyer Delph Carpenter, the commissioners chose to either ignore this information or challenge the credibility of the messenger. Ultimately, a review board of distinguished engineers and geologists would endorse LaRue's view that the water supply was insufficient, but by that time there was simply too much momentum for ratification of the 1922 compact and the authorization of the Boulder Canyon Project.

The Remarkable E. C. LaRue

LaRue was a westerner, born in Riverside, California, in 1879. He graduated from the University of California (Berkeley) College of Engineering in 1904 and went to work for the Reclamation Service, at that time a branch of the U.S. Geological Survey. When the Reclamation Service separated from the USGS in 1907, he stayed with the USGS. He worked initially in Salt Lake City until 1911, then out of Pasadena, California, for the Water Resources Branch until his separation from the USGS in 1927.[4] His sixteen years in Pasadena were spent studying the Colorado River. He would become well known for his comprehensive studies of the river, his river expeditions to survey its dam and reservoir sites, his opposition to the construction of Hoover Dam, and his view that "the amount of available water to be allocated was being substantially over estimated."[5]

After Water Supply Paper 395 was published in 1916, his efforts focused on river trips and field investigations of reservoir sites, reconstructing the flows of the river back to 1851, preparing a comprehensive development plan, and helping with an investigation of irrigated lands in Arizona. On the 1923 river expedition, which studied reservoir sites on the river from Lees Ferry to Needles, California, he was joined by Herman Stabler.[6] Stabler and LaRue were close associates, and Stabler was aware of LaRue's hydrologic concerns. LaRue's vision for the comprehensive development of the Colorado River, along with updated hydrology and geology information, was published in Water Supply Paper 556 in 1925.

LaRue supported damming the river at every possible point. His primary problem with Hoover Dam was that he felt that it should be part of a more comprehensive plan to essentially put dams on the Colorado River in a stair-step fashion all the way from the confluence of the Green and Colorado Rivers, in Utah, to Parker, Arizona. LaRue didn't like the singular focus on a large dam at the Black Canyon or Boulder Canyon sites at the southern end of the Colorado's canyon country and instead advocated for a series of coordinated dams that would maximize the amount of water available for consumptive use and power generation, while minimizing evaporative losses. Under his plan, the largest dam and reservoir would be located at the current Glen Canyon Dam location. He lost the argument over which dam should be first to the engineers at the Bureau of Reclamation and to the politicians from California, who preferred the Black or Boulder Canyon sites, but he did not go quietly.[7]

LaRue's Water Supply Paper 556 advanced the understanding of the hydrology of the Colorado River beyond the Fall-Davis report in two ways that undercut the compact framers' ideas about how much water they had to work with. First, he corrected for upstream water use, which they largely ignored. Second, he used innovative techniques to estimate flows in the late 1800s, in the time before river gauges, allowing a fuller accounting of both wet periods and past droughts.

The hydrologic methodology and findings presented in the 1916 and 1925 water supply papers are similar, except that the 1925 version has the benefit of a longer period of record.[8] To estimate past flows at Lee Ferry, LaRue adopted a different approach than the Fall-Davis report, the source document for most of the hydrologic information presented to the 1922 compact commissioners. Instead of reconstructing Lee Ferry flows based on the flow of the Colorado River at Yuma and then subtracting out the Gila River flows, LaRue reconstructed Lee Ferry from the upstream gauge records. He writes,

> The Yuma record is continuous from the year 1902, the date of the beginning of construction of the Yuma project of the Bureau of Reclamation. The chief difficulty in applying this record to the canyon section lies in the fact that there is a large and variable loss of water by evaporation from the stream channel, especially from the overflowed lands in the valleys between Yuma and Pierces Ferry. These lands are submerged and saturated by the annual summer floods. The area thus flooded varies from year to year, and the considerable amount of water passing into the dry, heated desert air by evaporation and transpiration from the rank growth of vegetation varies. It is impossible to estimate accurately the amount of water thus lost. A more accurate estimate of the water supply for the canyon section can be obtained from the records of the flow of the main stream and its tributaries in the upper basin.[9]

LaRue recognized Upper Basin water use was small in the 1890s, but then steadily grew through 1922, so he corrected past flows at Lee Ferry to estimate what the river's flows would have been with the higher 1922 level of depletions.[10] This was a critical step in understanding how much water the Colorado River really had. For the period of the late 1890s to 1922 LaRue calculated an average natural flow at Lee Ferry of 16.5 million acre-feet per year, about a million acre-feet per year less than the 17.5-million-acre-foot number used by Colorado's Carpenter and 1.6 million acre-feet less than the 18.1-million-acre-foot figure used by Arthur Powell Davis in later compact ratification discussions.[11]

LaRue and Stabler were the first to apply paleohydrologic techniques to estimate the flows of the Colorado River before the first river gauge measurements. This was critical. Droughts in the past, as recently as the late 1800s, did not show up in the river gauge measurements, but they mattered, because they could happen again. In his 1916 Water Supply Paper 395, LaRue argued that the elevation of the Great Salt Lake could be used "to show conclusively the periods of high, low, and average run-off." In Water Supply Paper 556, he expanded his use of the Great Salt Lake elevation data. As a closed basin, the Utah lake acted like a landscape-scale rain gauge—rising in years with more rain and snow and falling in years with less. As early as 1879, pioneering geologist G. K. Gilbert had used measurements of the Great Salt Lake's rise and fall as a crude climate proxy.[12] It was LaRue who first applied it to the critical question of estimating past river flows. First, he used the elevation data to estimate inflows to the Great Salt Lake. Then, under the theory that the Salt Lake watershed is adjacent to much of the Colorado River watershed and, therefore, the inflows to the lake should correlate well with Colorado River flows, he estimated Colorado River flows.[13] Using this technique, LaRue reconstructed the flow of the Colorado River at Lee Ferry back to 1851. However, he noted that accurate lake gauge-height records only went back to 1875. Prior to that the elevation of the lake was reconstructed from "the memories of some of the pioneers."[14] LaRue's early work closely matches modern estimates based on the latest tree ring research.[15]

The fit is not perfect. Before about 1880, the LaRue reconstruction significantly overstates the flows of the Colorado River. This might call into question the memories of some of the pioneers. However, after about 1880 the two reconstructions match each other fairly well; 1884 is the first year the ten-year running average is fully based on the more reliable post-1875 data. The important point is that LaRue's reconstruction accurately captured past droughts. It left no question that the two decades of data in the Fall-Davis report, used as the basis for the compact allocations, represented an unusually wet time.[16]

LaRue then adjusted his long-term reconstruction for future development in the Upper Basin. He concluded that the system did not always have enough water to meet the needs of both basins, reservoir evaporation and system losses, and any future burden to Mexico. If LaRue's analysis was correct, the basin's water developers were planning their projects on the back of an inadequate water supply.

Even before developing the Great Salt Lake technique, there was a second method available for measuring early flows on the Colorado River, but it was not clear how trustworthy it was. In the 1916 Water Supply Paper 395, LaRue included a chart of the "stage readings" at Yuma from April 1878 until 1902, a measurement of the height of the river. The Southern Pacific Railroad, with a river crossing at Yuma as well as the community's first municipal

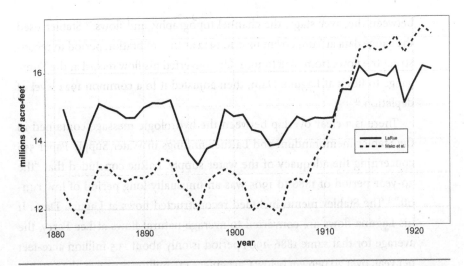

Figure 3 LaRue's estimated Colorado River flows compared to the modern tree ring record (ten-year moving average). Sources: LaRue data, WS556, modified to natural flows by the authors; tree ring data, "Colorado River at Lees Ferry, CO," TreeFlow, https://www.treeflow.info/content/colorado-r-lees-ferry-az-meko.

water system, needed to keep track of the river's elevation. Combined with measurements of a river channel's topography (the shape of the river bottom) and the velocity of its flow, stage readings provide a common method for estimating the volume of water moving through a river channel. But LaRue only had the height, not the other two measurements, and in his 1916 report he concluded that "no method has been devised whereby the early gauge heights can be used to determine the discharge."[17] He further explained that due to silt deposition in the river channel there is no reason for concluding that the average annual discharge for the period 1878–98 was less than that for the period 1899–1914.[18] In Water Supply Paper 556, LaRue makes no mention of the 1878–1901 Yuma river stage reading.

In the early 1920s, LaRue's USGS colleague, Herman Stabler, took up the challenge of finding a method to reconstruct the annual flows of the Colorado at Yuma based on the stage heights at the railroad bridge back to 1878. By that time, the Reclamation Service and USGS had collected more data on the flows and river channel at the Yuma gauge site. This additional data allowed Stabler to make a more detailed evaluation of the relationship between the river stage, the channel topography, and flows.[19] Stabler used the gauge data at Yuma from 1902 to 1922 as the calibration period to reconstruct the flows from 1878 to 1901. He converted his flow record at the Yuma gauge to flows at Laguna Dam, then adjusted it to a common 1922 level of depletion.[20]

There is a clear overlap between the hydrologic message contained in the Stabler memorandum and LaRue's findings in Water Supply Paper 556 concerning the adequacy of the water supply. LaRue concluded that "the 20-year period of 1886 to 1905 was an unusually long period of low runoff."[21] The Stabler memo included reconstructed flows at Laguna Dam. If his Laguna flows are converted to average natural flows at Lee Ferry, the average for that same 1886–1905 period is only about 13.5 million acre-feet per year, over 20 percent below Carpenter's 17.5 million acre-feet, the number that was central to the compact negotiations. Even though a twenty-year dry period of that magnitude should have been a major concern for the compact negotiators, Stabler avoided making any direct connection to the compact.

Despite receiving an October 4, 1923, letter from LaRue's supervisor, Nathan Grover, asking him to "please take especial care not to discuss or even mention controversial matters particularly as they relate to the 'pact'" (the 1922 compact), LaRue refused to avoid the compact.[22] He concluded, based on his Great Salt Lake reconstruction of Colorado River flows, that if the Upper Basin developed all the water to which the compact said it was entitled, a drought of the sort experienced in the late 1800s would mean there was not enough water in the system to meet the Upper Basin's legal delivery obligations. Not once but twice during the time covered by his study, flows dropped so low that, for five years at a time, the Upper Basin would not have been able to use its share of the river's water for its farms and cities *and* meet its 75-million-acre-foot obligation at Lee Ferry.[23] Thus, almost four years before the 1922 compact would be finally ratified by Congress, LaRue had published findings directly contradicting the critically important statements made during the ratification process by the compact commissioners from the Upper Basin states and Chairman Hoover that it would be highly unlikely or impossible for the Upper Basin to not meet the 75-million-acre-foot requirement of Article III(d).

Using the work of LaRue and Stabler, the available water supply, as measured by the average annual natural flow at Lee Ferry over the almost fifty-year period of the mid-1870s to 1922 was about 15 million acre-feet per year, significantly less than the 17.5 million acre-feet on which the compact's negotiators had relied and very close to today's modern value of 14.8 million acre-feet per year.

LaRue's final conclusion is "from these estimates it appears that when ultimate irrigation development is reached in the upper basin of Colorado River there will be an annual shortage of 5,000,000 acre-feet in the Lower Basin, an amount sufficient to irrigate 1,100,000 acres of land."[24] If we consider the situation on the river today, it can be argued that based on a 2016 level of development in the Upper Basin of about 4.5 million acre-feet per year, 3 million acre-feet less than LaRue's assumption of 7.5 million acre-feet, his conclusion was on point. Today, if the Upper Basin was only delivering 75 million acre-feet every ten years, the shortage to the Lower Basin would be about 2 million acre-feet per year.[25]

In simple terms, ninety years ago, a prominent USGS hydrologist using relatively modern hydrologic techniques predicted the situation we find ourselves in today with reasonable accuracy. The obvious question is why he was ignored?

Many of LaRue's problems were self-inflicted. Dianne Boyer and Robert Webb note that he was "self-centered and self-absorbed" and that he "pushed ahead, ignoring the subtle and not-so-subtle warnings about his personality that he received while on the river."[26] That personality trait was on full display in his December 1925 appearance before the Senate Committee on Irrigation, as he repeatedly interrupted and argued with the U.S. senators he was hoping to persuade.[27] Boyer and Webb note further that he "vociferously advocated his dam site and his water-development plan, without the official consent, support, or approval of the USGS or the Department of the Interior." They report that LaRue arrogantly closes a letter to G. E. P. Smith with the following: "I believe it would help clear the atmosphere if the politicians as well as the engineers would postpone discussion of the Colorado River projects until the engineering facts are made available." He was referring to his soon-to-be-published Water Supply Paper 556. Smith, a well-known irrigation engineer from the University of Arizona, was familiar with both LaRue's and Stabler's work and endorsed Stabler's conclusions in a 1925 report.[28]

LaRue's approach challenged his supervisors. They recognized both his brilliance and his limitations. Prior to the compact negotiations, he wrote Hoover offering his expertise to the negotiators.[29] He was not invited to do so, and there is no record that he was involved in the negotiations in any direct manner. Instead, Grover appears to have adopted a strategy of delivering some of LaRue's river knowledge through other more acceptable messengers. At the commission's field hearing in Phoenix, Grover and USGS engineer John Hoyt suggested a split of the river based on 65 percent of the available water to the Lower Basin and 35 percent to the Upper Basin.[30] The idea behind this proposal was that the more productive agricultural lands were in the Lower Basin. But they never brought up LaRue's concerns about the droughts of the 1800s and the risk that the river's flow was less than the compact's negotiators believed.

In "L'Affaire LaRue," Walter B. Langbein, a retired USGS hydrologist, is more sympathetic to LaRue. He notes LaRue's focus was on a single integrated basin-wide plan and that he emphasized that the demands for water would exceed the available supply and thus the water losses by evaporation should have been a serious and critical planning criteria.[31] LaRue's plea in front of the 1925 Senate hearing was that "all dams constructed on the lower Colorado River should conform to a comprehensive plan of development, which will provide maximum use of water resources and prevent an unnecessary waste of resources."[32] In contrast, Langbein presented LaRue's adversaries, the Bureau of Reclamation, the City of Los Angeles, and California's Senator Johnson and Representative Swing as singularly focused on a "one shot scheme" at Black Canyon or Boulder Canyon.[33] The Boulder Dam argument was advanced by major figures, including Hoover, who wanted a dam close to the power markets in Southern California, and Los Angeles's William Mulholland, who ridiculed the LaRue plan, which included a gravity aqueduct to Los Angeles.[34]

Langbein concludes that by 1926, the Boulder Canyon Project, with a high dam at Boulder Canyon or Black Canyon was "a settled issue." He also concludes that LaRue "deserves an honorable place in history." LaRue resigned from the USGS in 1927. He died of a heart attack in 1947 at the age of sixty-eight. At the time, he was working as a hydraulic engineer for the U.S. Army Corps of Engineers.[35] One of the ironies is that had LaRue's comprehensive plan been more favorably received, the bottom of the Grand Canyon might today be filled with "heel-to-toe" power dams. The other irony is that Bureau of Reclamation eventually built many of the other projects LaRue advocated, including Flaming Gorge, Glen Canyon and Davis (Bullhead) Dams. Others were proposed by Reclamation, but not built—most notably Marble Canyon and Bridge Canyon Dams in the Grand Canyon.

Carpenter and others blamed LaRue for Arizona's hardening opposition to the 1922 compact. Based on Carpenter's writings about LaRue, the state commissioners, except for Arizona, viewed him as an irritant intent on putting up obstacles to the ratification of the compact and a self-absorbed rival to Arthur Powell Davis and the other engineers at Reclamation.[36] They ignored or overlooked his hydrology expertise and conclusions, which could have helped them make better decisions, because they viewed his primary agenda as advocating for his plan for reservoirs with little regard for input from others. His standing with the other states was further diminished because they viewed him as an ally of Arizona governor Hunt in opposition to the compact.

In hindsight, LaRue's picture of the Colorado River hydrology as of the mid-1920s was far more comprehensive than that which Reclamation presented to the compact negotiators and the nation via the Fall-Davis report. When considered in tandem with the March 1924 report by Herman Stabler, by the early 1920s USGS scientists had developed a long-term record of flows on the Colorado River that is similar to our modern understanding. LaRue and Stabler both recognized and documented the now well-accepted fact that the Colorado River from about 1875 to 1905 was generally dry, similar to 1931 to the mid-60s. By comparison, 1906 to about 1930 was much wetter. Interestingly, if one looks at the entire period of about 1875–1922, the average annual natural flow at Lee Ferry is close to what we now accept as a long-term 1906–2016 average.[37]

It is true that unlike the period of 1900–1930, Reclamation had no specific stream flow data from the period of 1875 to the late 1890s with which they could conduct detailed hydrologic studies. However, it is clear they did have sufficient information upon which to justify a more conservative approach

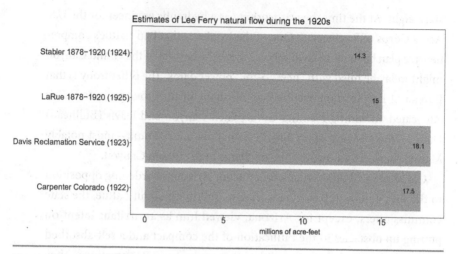

Figure 4 Comparison of early estimates of Colorado River flow.

to analyzing how much water was available to meet compact allocations. It would have at least allowed members of Congress, Interior department policy makers, and state water officials to have a more complete understanding of the consequences of the decisions they faced. Further, it might have set a precedent by which it was both acceptable and expected that Reclamation hydrologists, engineers, and economists could take a broader view of project feasibility analyses, including evaluating water project performance, unit costs and reliable yields, over extended dry periods. As we shall see in later chapters, it could be argued that it wouldn't be until the first decade of the twenty-first century before such an approach would be commonly used.

A side-by-side comparison of the four major estimates of the river's flow available to decision-makers by the mid-1920s illustrates the core issues. Boosters relied on estimates from Arthur Powell Davis and the Reclamation engineers, and those provided by Colorado's Delphus Carpenter. Their estimates of average annual flow, by ignoring the droughts of the late nineteenth century, were far larger than those provided by LaRue and Stabler—estimates that we now know are much closer to the Colorado's actual flow.

Might LaRue's and Stabler's hydrologic results, had they been made available to and accepted by the commissioners, have changed the outcome of the 1922 compact negotiations or its ratification?[38] We suggest that it depends on the timing of when the information was available. As previously mentioned, if the data had been available early in the compact negotiations, the negotiators might have split all of the water available without a surplus, and the Upper Basin's flow obligations at Lee Ferry and the ten-year period for providing the flow might have been different. We expect that a compact still would have been completed.

As to an impact on the process of ratification subsequent to the compact's 1922 negotiation, it had no impact, primarily because the commissioners did not accept its credibility. Had they accepted it, we believe Hoover and the six non-Arizona commissioners would still have pursued ratification. Ratification of the 1922 compact, even an imperfect one, was still better than the other options.

The only realistic alternatives to the ratification would have been either a congressional apportionment or, more likely, extended Supreme Court litigation for an equitable apportionment among the seven states. If the court chose to decide the fate of such a suit based on a basin-wide application of prior appropriation, as it did in the 1922 *Wyoming v. Colorado* case, the four Upper Basin states and Nevada would not have fared well. California might have been better off on paper. However, the length and complexity of such a case almost certainly would have delayed construction of Hoover Dam, the Colorado River Aqueduct, and the All-American Canal, its primary interests for a compact, until after World War II.

CHAPTER 6

The Sibert Report

A Lost Opportunity

"Sibert Report Jolts Boulder Dam Sponsors," the *Chicago Tribune* headline blared on December 4, 1928. The day before, an expert panel convened by Congress and headed by William Sibert, one of the nation's leading dam engineers, had delivered a report that, according to the *Tribune*'s Arthur Crawford, "caused elation among opponents of the project and jarred the confidence of its sponsors."[1]

Two great streams of the Colorado River's story—one of enthusiasm about the river's promise for the nation's flagging economy, the other skepticism on the part of a handful of scientists that the river had enough water to meet the enthusiasts' expansive dreams—were converging on the U.S. Congress in the days before the 1928 Christmas holiday. The confluence came in what in retrospect may have been the last great opportunity to bring hydrologic reality to bear on the dreams of the boosters—the *Report of the Colorado River Board on the Boulder Canyon Project*.

There in the pages of the Sibert board's report was a clear message. The nineteenth-century droughts first identified by LaRue and Stabler meant the Colorado River had less water than the boosters had imagined when they crafted the 1922 Colorado River Compact and the federal legislation now before the Congress to ratify the compact and launch construction of what would become the Hoover Dam. The report's math was inescapable. Once reservoir evaporation and water for Mexico were taken into consideration, any realistic effort to estimate the river's flow left too little water to meet the allocations carved out in the 1922 compact and about to be ratified by Congress in federal statute.

Whatever fears there may have been, though, were swept aside by a momentum that had become an inevitability. Less than three weeks later, pressed by a need to complete other work so members could get home for Christmas, Congress approved the Boulder Canyon Project Act, locking in the water allocations for good.[2] "Mighty Colorado Now Must Work for Man," headlined the *New York Times*. The river could be turned to irrigate half again as many acres of farmland as it was then doing, with enough water as well "to wet the throats, fill the bathtubs and water the lawns of citizens of Los Angeles and Pasadena, more than 200 miles away."[3]

The great lost opportunity of the board headed by William Sibert began in a last-ditch effort by critics of what had become known as "the Boulder Canyon Project" to throw up scientific roadblocks. As often happens in such controversies, it wasn't science they really cared about.[4] Instead, having largely lost the political battle, they hoped science might bail them out. The argument came to a head in a testy hearing before the House Rules Committee in May 1928. Led by Arizona congressman Lewis W. Douglas, opponents raised a welter of technical questions. Would earthquakes pose a risk to the dam? What of the engineering feasibility of the foundations for a dam that large? Did the dramatic and deadly failure of Los Angeles's St. Francis Dam two months before, which claimed more than four hundred lives, suggest problems with the very idea of building big dams? Was a dam that large needed? Might it actually increase the flood threat to downstream communities because of the risk of dam failure? Never did the critics raise what in hindsight seems the obvious question—was there enough water in the Colorado River to serve the projects for which the dam would be built?[5]

By the end of the month, Congress took a step toward answering the technical questions, authorizing one final technical report. A board of engineers would be tasked with advising federal decision-makers on "matters affecting the safety, the economic and engineering feasibility, and adequacy of the proposed structure and incidental works" to be built on the Colorado River.[6] The resulting report by what came to be known as the "Sibert board" undercut the critics' main arguments, concluding the gargantuan Hoover Dam could be safely built. But in taking seriously the charge to evaluable the dam's "economic . . . feasibility," the board also ventured one last time into the question of whether the Colorado River had enough water to do what was being asked of it.

Major General William Sibert was the epitome of the respected technical expert of his day—1884 West Point graduate, World War I infantry commander, and the engineer who oversaw construction of Gatun Dam on the Panama Canal.[7]

When Congress laid out the board's duties, its charge was expansive and the time line short: "to examine the proposed site of the dam . . . and review the plans and estimates made therefore, and to advise [the secretary of the interior] prior to December 1, 1928, as to matters affecting the safety, the economic and engineering feasibility, and adequacy of the proposed structure and incidental works."[8] The charge did not explicitly include reviewing the hydrologic assumptions associated with the Colorado River Compact, but that is what eventually happened in one of the great missed opportunities in the history of the development of the Colorado River.

The Sibert board did something no one else had done, conducting an independent review of the basic methodology used to estimate the inflow at the reservoir site (what would become Lake Mead). Key to that was the board's dissection of the Colorado River hydrology as understood by Reclamation at the time by focusing on an analysis of the Yuma gauge data. In so doing, they identified two fundamental flaws. First, the board's members concluded that mistakes in the use of data from the river's Yuma gauge led to an overestimate of the Colorado River's flow. Their straightforward conclusion: "In the opinion of the board the results of the Yuma gaugings are at least 10 percent too high."[9] Second, Sibert and his colleagues argued that the failure to consider data from the nineteenth century meant the risk of extended drought, visible in the flow reconstructions assembled by both Stabler and LaRue, had been ignored.

The Sibert report's conclusion about the river's available water supply stands in retrospect as one of the most important early declarations about the reality of the hydrology of the Colorado River: "One of the most important facts shown by these estimates is the existence of a long dry period in the Colorado River flow prior to 1906. This low period is clearly shown by an inspection of the Yuma gauge heights for that period. Further investigation has convinced this board that the flows of the Colorado River as determined by the gaugings from 1906 to 1927 are materially higher than the flow for the preceding 20 years, and that a long period of equally low flows must be expected to recur at any time."[10]

The Sibert board also found independent lines of evidence backing up this statement with gauging and precipitation records that preceded 1906 elsewhere in the basin, showing that the unusually high flows on which the compact had been based were not isolated to any one stretch of the river.

The Sibert report concluded its analysis with a call for the application of the precautionary principle, and one of the first pleas to come to terms with the river's variability: "The records of past performance of the Colorado River and of such other streams in this vicinity as seem pertinent furnish no basis for an exact estimate of long-past flow in the Colorado River. There is naturally considerable leeway in the interpretation of these data, and estimates based thereon may differ materially. The board however realizes that in determining economic feasibility of this project its estimates should be on the safe side." Rather than picking one number, the board suggested the planning consider a range, with flows available for future depletion ranging from 10 million acre-feet during drought periods as long as fifteen to twenty years, to high flows over similar time periods of 14.5 million acre-feet, with a long-term average somewhere in the middle.[11]

The Sibert panel's mission, and therefore its conclusions, were focused primarily on the economics of the dam itself. Would there be enough water to generate enough electricity to make the financing scheme being contemplated by Congress pencil out? This led inevitably, however, to a consideration of the Colorado River Compact itself, in language that was enthusiastically misinterpreted by boosters to imply that there really was enough water to meet their expansive plans. When existing Upper Basin irrigation was factored in, the river's natural flow as the river entered the Boulder Canyon Reservoir site is "about 15,000,000 acre-feet. This is the amount apportioned by the seven state compact at Lees Ferry."[12]

Those more familiar with the compact should have recognized that based on these numbers, the compact states were headed for problems: First, with only 15 million acre-feet of water available on the main stem, no surplus would be available to meet both a future obligation to Mexico and the future needs of both Arizona and California beyond the initial apportionments made by the 1922 compact. This involves the oft-forgotten Article III(f) of the compact, which presumed that there would be a surplus available above and beyond the specific allocations to the Upper and Lower Basins. Such a surplus, which would be available in the future, after 1963, was politically essential to both the negotiations of the 1922 compact and, ultimately, the six-state ratification process provided for in the Boulder Canyon Project Act. The act required California to limit itself to 4.4 million acre-feet per year plus half of the available surplus as a prerequisite to the six-state path. By 1929, California already had plans in place to use more than 4.4 million acre-feet per year. The perceived availability of a surplus was politically required before the California legislature would consider this limitation.

Additionally, the Sibert board's conclusion that a period of low flow could last as long as twenty years is inconsistent with the assumption in Article III(d) that a ten-year period is sufficient to "allow equalization between years of high flow and low flow," as Hoover and Carpenter testified.

Finally, a simple mass balance shows that under the Sibert hydrologic assumptions the Upper Basin would not have 7.5 million acre-feet available to it for consumptive use. The Sibert board concluded that 15 million acre-feet was available at Black Canyon, not at Lee Ferry. By this time, it was well understood that Lower Basin tributary inflow between Lee Ferry and the Black Canyon dam site was in the range of 800,000 to 1 million acre-feet per year. This meant the available natural flow at Lee Ferry was no more than 14.2 million acre-feet per year. A 7.5-million-acre-foot-per-year average annual obligation at Lee Ferry under Article III(d) only left 6.7 million acre-feet available for consumptive uses in the Upper Basin. Achieving even the 6.7 million acre-feet of consumptive use would require both no Upper Basin obligation to Mexico, which conflicts with Article III(c), and the construction of a large amount of upstream storage above Lees Ferry. This storage ultimately was built, but as of the 1920s, it was not guaranteed.

The report's numbers should have come as no surprise. They were based on information that was available by the early to mid-1920s. As we have seen, the Stabler memo referred to by the Sibert board was included in the congressional testimony submitted by Secretary of the Interior Hubert Work to the Committee on Irrigation and Reclamation of the House of Representatives in March 1924.

Clearly, the 1922 compact commissioners and their advisors should have been aware of the potential hydrologic problems well before the compact became effective. An obvious question is why the commissioners from the basin states and their advisors did not take the opportunity to revisit the 1922 compact hydrology or call into question their state ratifications.

There are two reasons. First, many of the commissioners and their advisors believed the Sibert board findings were too conservative. As project opponents in Wyoming tried in early 1929 one last time to resurrect the concerns raised in 1923 by Nellis Corthell, the Sibert board's findings were, through misunderstanding of the math, used to provide reassurance rather than raise questions. Wyoming governor Frank Emerson, who in 1922 was the Wyoming commissioner and in 1926 was elected governor, told Utah's governor George Dern, "I do not hesitate in giving my continuing conviction that the Upper States are safe within all reasonable limits. . . . My own conclusion has been supported by several other engineers well versed in water supply matters and possibly better qualified through virtue of their practical experience to pass upon such questions than . . . the engineers rendering the Sibert report."[13] The second reason was that even with those caveats, the "conservative" Sibert calculations showed a 15-million-acre-foot-per-year river, enough to meet Upper and Lower Basin needs, Wyoming congressman Charles Winter suggested in a February 14, 1929, letter to the Big Piney Examiner.[14] As we have seen, that is a misinterpretation of Sibert's findings, but in Wyoming it was enough to overcome reservations that had lingered since Corthell's attempt to raise the question of the river's hydrology six years before.

What happened in Wyoming illustrates the political reality that was the most important reason the Sibert board's concerns about the available water supply had so little impact. There was simply too much momentum in Congress for passing the Boulder Canyon Project Act and implementing a six-state ratification of the 1922 compact. The Senate passed its version of the bill on December 14, 1928, just eleven days after the report was transmitted to the secretary of the interior. Then, after the House concurred with the Senate amendments on December 18, President Coolidge signed the Boulder Canyon Project Act on December 21. With California's March 4, 1929, approval of the California Limitation Act—a requirement for a six-state compact under the act—the last pieces were in place for the great engineering work to proceed.

CHAPTER 7

Boulder Canyon Project Act

It was a triumphant moment when Herbert Hoover was called upon as president in June 1929 to sign the proclamation formally bringing the terms of the Boulder Canyon Project Act into force. "I have a particular interest in its consummation," he told reporters, "not only because of its great intrinsic importance but because I was the chairman of the Colorado River Commission that formulated the compact. . . . It is the final settlement of disputes that have extended over 25 years and which have estopped the development of the river. The difficulties over the respective water rights of the different States have served to prevent development in a large way for nearly a quarter of a century."[1]

Now that the Boulder Canyon Project Act was effective, three of the four basic pillars of the Law of the Colorado River were in place: the prior appropriation doctrine, the 1922 Colorado River Compact, and the 1928 Boulder Canyon Project Act. The fourth pillar, an international water treaty with Mexico, would follow in fifteen years.

The prior appropriation doctrine is not often included in discussions of the Law of the River, but it was crucial, because without the threat of the interstate or basin-wide application of prior appropriation, there would have been no reason for the Upper Basin states to even consider a compact. The Colorado River Compact is primarily a political document that shapes the appropriation doctrine into an acceptable interstate doctrine among states that had unequal needs and vastly different economic potentials but, through the vehicle of the United States Senate, almost equal political power. One of its fundamental purposes is to limit how appropriators in one basin could impact appropriators in the other basin. The compact cleared the way for the political coalitions that would be needed for the development of the Colorado River Basin, a social contract between the Upper and Lower Basin states. Because the compact subordinated hydroelectric power to irrigation and domestic uses, the agreement effectively shut the door on large private power interests and opened the door to development that would be dominated by the federal government for irrigation districts and the region's large municipalities. The states would play a large political and, in some cases, regulatory role but invested little of their own money in the development of the basin's major water projects.

Hoover spoke with pride about the compact, a historic achievement not only because of its importance in the development of the West, but also in the relations among states working to find ways to act as equals under United States federalism to collectively solve a problem:

> It is the most extensive action ever taken by a group of States under the provisions of the Constitution permitting compacts between States. The only instances hitherto were mostly minor compacts between two States on boundary questions except the one case of the New York Port Authority, which was of first importance, but is [a] compact between two States. This compact is, however, an agreement between seven States, and represents the most important action ever taken in that fashion under the Constitution. It opens the avenue for some hope of the settlement of other regional questions as between the States rather than the imposition of these problems on the Federal Government.[2]

But Hoover's comments foreshadowed problems. "There is only one point still left open," he told reporters, "and that is the relation of Arizona to the compact. I am in hope that Arizona and California may compose their mutual problems which have hitherto prevented Arizona from joining in the compact. With Arizona in, the whole basin will have settled their major question of water rights for all time." Arizona would not be "in" for decades to come, creating festering problems that undercut Hoover's enthusiasm for the pioneering agreement.[3]

When Congress had approved the act the previous winter, it was only the beginning of a tortuous political and legal process. Rather than authorizing a blitz of infrastructure construction outright, the legislation required major conditions be met before projects could proceed. The most formidable was a requirement that the deal had to, again, go back to the states for final approval. It required either (1) ratification by all seven states of the Colorado River Compact or (2) if seven states should have failed to ratify within six months, six states had to ratify it provided the six states waived Article XI of the compact, which calls for ratification by all seven states; one of the states had to be California; and the California legislature had to pass an act limiting itself to 4.4 million acre-feet per year of consumptive use per year plus one-half of the available surplus. Additionally, the 1928 act required that before any money could be expended on the Boulder Canyon Dam and power plant or the All-American Canal, the secretary of the interior had to enter into the appropriate repayment contracts.

The Act's Key Provisions

In providing the escape hatch of a six-state ratification process, the act embedded one of Colorado River management's biggest problems—uncertainty over Arizona's share of water. What Hoover had hoped would be a broad social contract of equals among the seven states became a contentious battleground for decades to come, in ways that remain unresolved to this day. But in more subtle ways, the act also embedded other problems that rest deeper in the relationship between the data used to underpin the deal and the river's actual hydrology.

The most important of these embedded problems was the act's suggested math surrounding allocation of water among the three States of the Lower Division—California, Arizona, and Nevada. It specifically referred to the 7.5 million acre-feet apportioned to the Lower Basin under Article III(a) of the Colorado River Compact, suggesting an apportionment of 300,000 acre-feet to Nevada, 2.8 million acre-feet to Arizona, and in theory limiting California to 4.4 million acre-feet. If that was all the act did, it might have been sufficient to bring the system into balance with hydrologic reality. But in a provision that came back to haunt basin water managers in the decades to come, the act authorized Arizona and California to annually use one-half of the surplus water not apportioned by the Colorado River Compact. In ignoring the cautions of LaRue, Stabler, and Sibert, that provision allowed California to live with the belief that it might in practice be able to use 5 million acre-feet or more per year in perpetuity, a provision that did not collide with hydrologic reality until after the turn of the twenty-first century. Representatives of the Upper Basin states believed that including it in the 1928 act would ease their concerns over Arizona's failure to ratify the compact. The sponsor of the amendment, Senator Key Pittman of Nevada, strenuously argued that this provision was a "suggestion," not a requirement of Congress.[4]

The act also settled by act of Congress something that had been left unresolved in the state-to-state negotiations that led to the compact, by granting Arizona the exclusive beneficial use of the Gila River and its tributaries within Arizona, and noting that, except for return flows, the Gila River would never be subject to any "diminution" by water which may have to be delivered to Mexico under Article III(c) of the Colorado River Compact.

The problems with these provisions are significant. First, the 1928 act makes no mention of Article III(b)'s allocation of the extra million acre-feet to the Lower Basin, even though Hoover had testified in 1923 that the supply for III(b) was "both the mainstream or from any of its tributaries."[5] Congress's failure to even mention Article III(b) would lead to confusion and litigation, and questions concerning Article III(b) remain unresolved today. Second, the Lower Basin as defined by the Colorado River Compact includes the upper Gila and Virgin Rivers—watersheds in New Mexico and Utah that drain into the Colorado River below Lee Ferry. Any Lower Basin compact would have to include Utah and New Mexico as signatories. Third, in limiting California to 4.4 million acre-feet, Section 4(a) appears to define consumptive use as "diversions less returns to the river," ignoring evaporation from the large reservoir authorized by the 1928 act itself—another major issue that remains unresolved.

As we shall see in later chapters, a dispute between Arizona and California over how much water was legally available to Arizona would force the U.S. Supreme Court to take an expansive view of the act's provisions.

The reason Section 4(a) became much more than a suggestion is that Section 5 requires all contracts to conform to 4(a) and that "no person shall have or be entitled to have the use for any purpose of the waters stored (in Lake Mead) . . . except by contract." With this provision the secretary of the interior would become the water master for all water uses on the main stem of the Colorado River in and below Lake Mead. In theory, had the Lower Basin been successful in negotiating a compact, the congressional approval would have likely been implemented in a manner that replaced Section 4(a) with the appropriate compact provisions.

Section 15 of the act directed the secretary of the interior to formulate "a comprehensive scheme of control and the improvement and utilization of the water of the Colorado River and its tributaries." This study, which would take almost twenty years to complete, would set the stage for most of the river's post–World War II development, reaching upstream of Hoover Dam and the canyon country of Utah and Arizona into the headwaters states of New Mexico, Colorado, and Wyoming, to create a framework for the network of federally funded dams and diversions that we see on the river today.

The final critical provision is Section 19, which gave congressional sanction for the basin states to "negotiate and enter into compacts or agreements, supplemental to and in conformity with the Colorado River Compact." It also required that a representative of the United States, appointed by the president, participate in the negotiations and any such compact or agreement be approved by the legislature of each participating state and by the U.S. Congress. The five states with Upper Basin lands used this authority in the late 1940s to negotiate the Upper Colorado River Basin Compact.

Ratification of the Six-State Compact

The seven-versus-six-state options included in the act were Congress's attempt to deal with the reality that Arizona might not approve the agreement. With all eyes on Arizona, there was a pause in early 1929 to see what the desert state would do. But as it became clear Arizona would not ratify the compact, the other states began to act. California, with the most at stake both in terms of the flooding risk in the Imperial Valley and the economic benefits of water development, passed the necessary legislation. California governor C. C. Young signed the California Limitation Act on March 4, 1929, but the drama was not over. Opposition to a six-state ratification surfaced in Utah, led by Senator Reed Smoot and Representative Elmer Leatherwood, both of whom had opposed the Boulder Canyon Project Act in Congress. Leatherwood's questions about earthquake risk at the Boulder and Black Canyon dam sites contributed to the congressional request that led to the Sibert board's report. Utah's opposition was not really about seismic risk at all but rather a fear that, in assigning "water master" responsibilities to the federal government on the Lower Colorado, the bill set a precedent for federal control of water within a state's boundaries. The Sibert board's report gave Utah opponents an added argument—a fear about the adequacy of the river's water supply. But the opposition represented a minority, and the Utah legislature easily passed the necessary legislation allowing a six-state compact.[6]

As Arizona's stubborn opposition demonstrated, the compact was far from a perfect political strategy. Its negotiators used a flawed understanding of both the magnitude of the river's flows and the variability of those flows. This understanding allowed them to proceed with the design of a compact that could be sold to the states and Congress. There was no desire to question the long-term yield of the river, and when inconvenient information surfaced, as we have seen in the cases of LaRue, Stabler, and Sibert, it was ignored or dismissed. It has been common in recent decades to hear water lawyers in Colorado make the argument that at some point in the future Colorado and its fellow Upper Basin states may have to challenge the legality of the 1922 compact on the theory that it was a contract based upon a mutual mistake of facts and therefore void.[7] We would suggest that the more compelling argument may be made that instead of a mutual mistake of facts there was a mutual decision to avoid and ignore hydrologic facts, especially when those facts were inconvenient to the goals of the compact negotiators.

The 1928 Boulder Canyon Project Act is considered almost an equal pillar to the 1922 compact because of its profound impact on both the physical, legal, and political fate of the river. In addition to authorizing the construction and operation of what came to be called Hoover Dam and the All-American Canal, it provided for a six-state ratification of the compact, it authorized future compacts, and it authorized a comprehensive study of the development of the Colorado River.

Equally important was the message the act conveyed to the states and public. First, the federal government could and would accomplish great engineering achievements, "taming" the heretofore unruly and dangerous Colorado River. Construction of Hoover Dam was one of this country's transformational achievements, remaking not only the Colorado River Basin but the nation itself, helping launch the American century.[8]

Second, while the compact negotiators may have viewed the basin states and federal government as "equal sovereigns," Congress made it clear the federal government was "more equal." It rejected the compact conclusion that the river was no longer navigable. By requiring contracts for all users of Lake Mead water, it made the secretary of the interior the water master on the main stem from Lake Mead to the Mexican border. And, whether by design or default (the default being the failure of the Lower Basin states to settle their own differences by negotiating a Lower Basin compact), it imposed a congressional apportionment on waters of the lower main stem. Finally, through the six-state ratification process, both Congress and the states made it clear that they would not allow the intransigence of one state to block regional progress. In this case it was Arizona, but in later decades, other states would be run over as well.

CHAPTER 8

More Marketing Than Science

"I view this great Colorado River project as a portent of new thinking and of new methods for the advance of our country," Interior secretary Ray Lyman Wilbur said at the ceremony formally launching construction of Hoover Dam. "Here we will listen to the voice of the expert who knows his business and is the only safe guide of democracy in its ever-present fight with the forces of nature."[1]

The challenge of "this great Colorado River project" was to now turn the legal structures built in the 1920s into actual dams, power plants, and canals. That required wrestling for the first time with a new and complex set of details—how much power could the new dam produce, a question which translated directly into the all-important question of how much revenue its operation would generate to pay back the costs of its construction. That, in turn, depended on the central question—how much water was there in the Colorado River? It was an unprecedented challenge to the use of expertise in executing the task of governance.

But which expert to choose? By September 1930, as Wilbur launched the greatest public works project the nation had ever seen, one expert, E. C. LaRue, had left the stage, resigned from the U.S. Geological Survey and bitter in his departure from the nation's work on the Colorado River. No longer would the nation "listen to the voice of the expert" who had warned there was not enough water to carry out the Bureau of Reclamation's grand plan. In his place, now, came Erdman Bruno (E. B.) Debler.

Gone also was the science agency—the U.S. Geological Survey. The call for expertise was now entrusted entirely to the agency with an incentive to build the dam. Also gone was the limited federal role Hoover had championed during the ratification of the 1922 compact. While the Fall-Davis report and Arthur Powell Davis's advisory role in the compact negotiations had set the stage, the Bureau of Reclamation was now in the position to benefit as both evaluator and promoter of projects. Together with the local boosters, including state water agencies, and elected politicians, the three would become known as the "iron triangle."[2] All three would benefit. Boosters got federal money for projects that would foster growth in their area, the politicians got public support, and the Bureau of Reclamation got appropriations to build an empire of water projects throughout the West. If this arrangement was a conflict of interest that discouraged the application of the sound and conservative science the Sibert board had recommended, no one seemed to notice.

Debler had begun his career with the Reclamation Service in 1914. By the late 1920s, he was one of its top Colorado River hydrology experts, where he had been a strong and very public advocate for federal construction of a dam at Boulder Canyon.[3] In the Boulder Canyon Project Act, Debler and his colleagues got what they wanted. The states and the politicians had done their job, approving the 1922 compact and federal legislation authorizing the great dam's construction. The construction engineering community had also done its job, advancing concrete technology and the design of high concrete dams to the point that there was no longer any question that a 700-foot-high dam was both feasible and safe. Debler's job now was essential. Unless the project could produce enough reliable and economically competitive hydroelectric power to repay the federal treasury for its cost, there would be no project. He fully understood that the amount of power produced was dependent upon the water supply. His job was to show that the project would produce enough power to succeed.

As Debler took up his task, the basin was about to be hammered by two major shocks, vast forces far beyond the project itself that would have profound implications. The first was economic, the Great Depression, which created a sense of urgency around the project as both an engine of job creation and a demonstration that the battered nation could still do great things. The second was hydrologic, the drought of the thirties, the first opportunity for a reckoning with the reality that there was in fact far less water available than the planners of "this great Colorado River project" had reckoned.

The Hoover Dam Power Contracts

The development of the Hoover Dam power contracts was one of the most important water policy processes on the Colorado River in the first half of the twentieth century. Removed from the politics of Congress and interstate negotiations, it brought together the federal government in the form of the Interior department and the people who would buy the dam's electricity. Hoover Dam and the legal superstructure developed to enable its construction were always first and foremost about water—about overcoming the "menace" of lower Colorado River flooding and turning water out of the river's channel to irrigate the desert. But those steps depended crucially on the generation of hydroelectric power. It was by extracting electricity from water falling through the river's dams that the whole scheme would be paid for. Section 4 of the Boulder Canyon Project Act made that explicit—no money was to be spent on the dam's construction until sufficient power sales contracts were in place to repay the U.S. Treasury the project's cost.[4]

To develop those contracts, Interior needed a good understanding of the costs of the project and the amount of electrical energy it would produce over the fifty-year repayment period. Interior needed to recover an estimated $200 million in power sales over the fifty-year period, a staggering sum for the early 1930s.[5] To estimate the amount of energy produced, the secretary had to estimate the quantity of water available over the fifty-year repayment period. To make this estimate, Interior turned to E. B. Debler.

The Debler study gave the Bureau of Reclamation an opportunity to revisit the Sibert board's findings, which had called into question Reclamation's water supply assumptions. Would Debler's hydrology consider the period from the late 1870s to about 1900, a period that was by this time well known to have been much drier than 1900–1930? Debler was certainly aware of the work of previous scientists, including LaRue and Herman Stabler, who had argued for considering the importance of nineteenth-century droughts. Debler served on the 1924 engineering committee that prepared the report that included Stabler's memo outlining the issue. Debler and his fellow Reclamation hydrologists should have been aware that river flows from the 1870s through the early 1900s were lower than the first three decades of the 1900s. As we have seen, in understanding the long-term flow of the Colorado, the period of record one uses for analysis makes all the difference.

But Debler passed up this opportunity to reconsider the hydrology in light of the critical decision it was meant to inform. The Debler study was basically an update of the Fall-Davis report hydrology, adding 1921–28 to the original study period used by the Fall-Davis report. As luck would have it, the years he added were wet, while the early years he continued to ignore were dry.[6] While choosing the period of record to use in deriving estimated flows of the Colorado River had long been important, as LaRue's attempt in the 1920s to consider nineteenth-century droughts showed, Debler's analysis appears to be the first time anyone explicitly constrained the period of record used to derive a preferred outcome.

As a result, Debler's estimate of the water available at the dam site, an average natural inflow into Lake Mead of 18.5 million acre-feet per year (17.5 million acre-feet at Lee Ferry), was extremely optimistic. Perhaps there was a deliberate policy decision to ignore LaRue, Stabler, and Sibert and by doing so ignore evidence of the pre-1900 droughts. Or perhaps Debler simply didn't want to go back before he had reliable streamflow data derived from actual gauges. Whether it was intentional or not, the optimistic hydrology helped the secretary of the interior market the Boulder Canyon Project's hydroelectric power at a "competitive" rate. For the same monetary investment in the dam and power generation facilities, a projection of 20 percent more available water reduced the cost of each kilowatt-hour by roughly 20 percent.[7]

With the Debler hydrology report in hand, Interior and three California entities—Metropolitan Water District of Southern California (MWD), the City of Los Angeles, and Southern California Edison—began power contract negotiations. For Interior, the negotiations were led by Northcutt (Mike) Ely, then assistant to the secretary of the interior, later co-author of the 1933 and 1948 editions of the Hoover Dam Documents and ultimately Chief Counsel for the Colorado River Board of California. Because neither Arizona nor Nevada had an immediate need for any project power, California interests agreed to pay for 100 percent of the firm energy but agreed that the secretary could take back up to 36 percent, 18 percent each for Arizona and Nevada, any time during the fifty-year contract period.[8] MWD received the largest single share of project power, initially 50 percent, which it would use to pump water from the Colorado River to the booming service area on the coast.

With the preliminary power contracts in place, Congress approved Secretary Wilbur's request for the initial $10.66 million appropriation. President Hoover signed the legislation July 3, 1930. On July 7, Wilbur sent a simple telegram to Reclamation commissioner Elwood Mead: "Sir: You are directed to commence construction on Boulder Dam to-day."[9] On September 17, 1930, Wilbur also directed Commissioner Mead to formally call it the Hoover Dam.[10]

Hydrology as a Marketing Tool

Debler's optimistic hydrology might be more accurately described as successful marketing strategy rather than science-based hydrologic analysis. To make the numbers work, Debler not only had to ensure there was, on paper, plenty of water to generate power at Hoover Dam and thus pay for the project. He also needed to ensure he did that while preserving the ambitions of the Upper Basin states to develop the river's potential for themselves.

History would show Debler's inflow hydrology was wildly optimistic.[11] Here again, he exposes the important distinction between natural flow— what would have been in the river absent upstream dams and diversions— and actual flow. Debler's assumption that by 1988, actual inflow to Lake Mead would average 11.9 million acre-feet per year is not that much higher than the estimated actual average inflow of 10.9 million acre-feet over the period of 1937–2001.[12] To make that work while still leaving a cushion for substantial irrigation development in the Upper Basin, he assumed that water use in the Upper Basin would ramp up from 2.76 million acre-feet per year in 1928 to 6.6 million acre-feet per year by 1988. That left him with the best of both worlds—an average inflow to Lake Mead over the 1938–88 period of 13.5 million acre-feet per year, plus plenty of water for irrigation and municipal water use in Wyoming, Colorado, Utah, and New Mexico. That cushion, enabled yet again by overly optimistic hydrologic math, removed a potential political stumbling block by keeping all the states of the Colorado River Basin happy, at least on paper.

That actual historical inflows were so close to Debler's projection—he projected 11.9 million acre-feet by the late 1980s, the actual inflow was about 10.9 million acre-feet—is a result of offsetting errors. First, actual flow in the river was far less than Debler projected. Second, Upper Basin development was much slower than Debler (and the Upper Basin optimists) had planned.

There are several reasons for the second point. First, the available flat river valley land that could easily be served by gravity ditches was fully developed by the early 1920s. Second, some of the reductions in irrigation depletions beginning in the 1930s were due to limited water availability. Up until the early 1930s, conditions were generally wet. In in the 1930s, conditions turned dry, and the combination of a lack of local storage and low physical water availability limited depletions. This is one of the critical characteristics of Upper Colorado River Basin water use. Without reservoir storage upstream, many users in dry years have no water to use, so rather than draining reservoirs, as is done in the Lower Basin during droughts, water use simply goes down. The final factor slowing Upper Basin development was the Great Depression, which beginning in the 1930s drove many small farmers and ranchers out of business and therefore left lands previously irrigated unused.

Debler thus continued trends that began with the compact negotiations and ratification in the 1920s, and which have continued throughout most of the history of the development of the Colorado River. In addition to finding numbers for both hydrology and water use that were politically acceptable and allowed projects to move forward, where there was hydrologic information available that would call to question the hydrologic or economic performance of a project, it was ignored.

The Hoover Dam Water Contracts

With the power contracts at least tentatively in place, the negotiators turned to contracts for water delivery. It is here that Debler's overestimation of the available supply of water left its most profound mark on the development of the Colorado River Basin, locking into place mistaken assumptions about the water supply that plague river management to this day. The roots of those assumptions are deep, but they finally surfaced in the process through which the secretary of the interior signed contracts with California water users for more than 5.3 million acre-feet per year of water. While the Boulder Canyon Project Act had only ensured California 4.4 million acre-feet per year, the notion of "surplus" allowed an overreach on which California water users came to depend but which by the twenty-first century was no longer reliably there. It was here, in the Hoover Dam water contracts, that the error became a part of practical on-the-ground water operations.

The 1928 act required all users of Lake Mead water to have a contract with the secretary of the interior. The intent was that revenues from both water and power would be used to repay the capital costs of the project, but it was always understood that power revenues would provide the bulk of the repayment.[13] The water and power contracts were intertwined through the Metropolitan Water District (MWD) of Southern California. MWD was acquiring up to 50 percent of the initial Hoover power to pump water from the Colorado River to its service area on the coast. However, it would not sign a final firm power contract without also obtaining a water contract, because without water it would have no use for the power.[14] A water contract with the Imperial Irrigation District was required before Reclamation could begin work on the All-American Canal.[15]

California water users negotiated an agreement allocating 3.85 million acre-feet to four agricultural districts, most notably the Imperial and Coachella Valley Irrigation Districts, and 550,000 acre-feet to the Metropolitan Water District. To that top-priority allocation of 4.4 million acre-feet, the contracts added another 550,000 acre-feet of lower priority water to MWD from the presumed available "surplus," with similar allocations of lower priority surplus water to San Diego and the agricultural districts. In all, the agreements totaled 5.362 million acre-feet of water. The "surplus" was crucial. Without access to more than a million acre-feet of water per year, MWD's Colorado River Aqueduct might not have made economic sense. The priority system of distributing California's share on paper set up a system for reducing deliveries to the lower priority users if there was insufficient water available beyond the basic allocation of 4.4 million. But as a practical matter, it allowed cities and farms to become dependent on excess water that was, in the long run, an unreliable supply.

On August 18, 1931, seven California water agencies signed the allocation agreement, referred to as the "seven-party agreement," paving the way for the secretary of the interior to enter into individual water contracts.

The first two water contracts were with the Metropolitan Water District and the Imperial Irrigation District. These contracts related to the construction on the two major projects authorized by the 1928 act, Hoover Dam and the All-American Canal. The MWD water and power contracts set the stage for the third major California project of the 1930s, the Colorado River Aqueduct.[16] Parker Dam, which holds back Lake Havasu, is an essential feature of the aqueduct. It was built by the Bureau of Reclamation but funded by MWD.[17]

By the end of 1934, the California parties had entered into contracts totaling the full 5.362 million acre-feet per year allocated in the seven-party agreement. The contracts made clear the supply was only as "firm" as the water supply to California under the 1922 compact and 1928 act. They were, in other words, dependent on surplus water beyond the basic 4.4 million acre-feet of water. Second, the contracts included language that required the United States to absorb all transit losses, evaporation losses and intentional and unintentional overdeliveries.[18] Again, in implementing the projects, their developers ignored hydrologic reality. Water evaporates from reservoirs and is lost as it moves down river channels, and that is every bit as important as water diverted from a stream for human use.

In allocating water, the contracts maximized the diversion for human use, using the entire amount allocated under the 1922 compact and 1928 act, while ignoring the fact that system losses would create a deficit. In the twenty-first century, that has become a significant problem. It takes three days for water released from Parker Dam to make it to Imperial Dam. If water ordered by a user on the lower river turns out not to be needed due to a rainstorm or system interruption, the ordered water flows past the diversion point (and normally ends up as extra water for Mexico) and under this contract language is not considered a delivery.[19] The other problem is that this language has led to a situation in the Lower Basin where reservoir evaporation, a significant man-made consumptive use of about a million acre-feet per year, is not apportioned to individual states.[20]

An obvious question is why the United States entered into contracts totaling 5.362 million acre-feet per year when the 1928 act required California to limit itself to 4.4 million acre-feet per year. There are several good reasons. First, the limit under the 1928 act and the California Limitation Act is 4.4 million acre-feet *plus* one-half of the surplus. Under the assumption that the main stem could support the allocation of 10.5 million acre-feet of water for contracting and delivery to Mexico, there was a surplus.

The second is the economic reality of the day. As the authors of the 1948 *Hoover Dam Documents*, Northcutt Ely and Ray Lyman Wilbur, bluntly noted, "It became necessary to proceed with water contracts with those agencies which were required by the Project Act (1928 act) to assume the repayment obligation for the project without further delay, or else indefinitely postpone the project."[21] Given the country's grave economic and political conditions in 1930 and 1931 and President Hoover's personal involvement in Colorado River matters, the prospect of Secretary Wilbur postponing the construction of the Boulder Canyon Project was unimaginable.

Wilbur offered a draft water contract to Arizona as well. The proposed contract would provide Arizona with 2.8 million acre-feet of water from Hoover Dam.[22] The contract was carefully written to be neutral with respect to the disputed compact issues between Arizona and California. The contract also provided Arizona with the right to 18 percent of the Hoover Dam power and 50 percent of the power that would be produced at Parker Dam. The Parker Dam power was offered at no capital cost to Arizona.[23] The proposed contract also required Arizona not to interfere with the development of projects on the lower river. There were several meetings, but no progress was made. Arizona had already chosen a path of litigation. In the 1940s, Arizona would sign a contract for 2.8 million acre-feet per year and Nevada contracts totaling 300,000 acre-feet per year.[24]

Thus, by 1944, the secretary of the interior had entered into water contracts totaling 8.462 million acre-feet per year from Hoover Dam. The water and power contracts with the California agencies were the keystones that built Hoover Dam. Because they had no near-term need for either water or power, Nevada and Arizona were little more than bystanders. Had Debler concluded that the natural inflow to Lake Mead was just a million acre-feet per year less than his 18.5 million acre-feet, the river as we know it today might be completely different. Indeed, he easily could have concluded that the available water was insufficient to support contracting for more than 4.4 million acre-feet of water to the California agencies. This might have made a 1.2-million-acre-foot-per-year Colorado River Aqueduct infeasible. It would almost certainly have made Hoover Dam's power more difficult to market. Did Debler and his Reclamation colleagues understand this dilemma? We believe the answer is yes, and, thus, the hydrology report was designed to make the Boulder Canyon Project look as good as possible.

The contracting of more than 4.4 million acre-feet of Lake Mead water to the California agencies became a major legal issue almost immediately. It would trigger decades of litigation and political skirmishes. However, even after what would be a shockingly dry decade of the 1930s, it was still not a practical problem. At the time, with full development still decades away, there was plenty of water to meet everyone's needs. It would be decades before Arizona and Nevada would have the projects in place and the demand to take anywhere close to the amount of water contemplated by their contracts. As (bad) luck would have it, the full development of Arizona's and Nevada's allocations of main stem water would coincide with a drought period in the early 2000s that would be far drier than the 1930s.

In 1930, presidential mediator William Donovan proposed a Lower Basin compact based on Debler's same optimistic hydrology. This would have given California 4.9 million acre-feet of apportioned main stem water plus half the surplus. Arizona would get full use of its tributaries, including the Gila River, a total use then believed to be 3 million acre-feet per year, plus 3.3 million acre-feet of apportioned main stem water, plus one-half of the surplus. Nevada would get 300,000 acre-feet of main stem water. Both Arizona and the Upper Basin states signaled a willingness to accept the Donovan Proposal. The willingness of the upper states to accept the Donovan suggestion is likely due to Carpenter's fear that Arizona could prevail in the U.S. Supreme Court. It's also evidence that they were comforted by the optimistic hydrology.

Had this proposal been accepted by the California, it might have at least delayed a number of political disputes and several rounds of interstate

SIDEBAR
Donovan's Compact That Never Was

One of the Colorado River Basin's great "what-ifs" is Colonel William Donovan's proposed Lower Basin compact of 1930.[25]

Appointed in 1929 by President Herbert Hoover to try to broker a deal among Nevada, Arizona, and California, Colonel Donovan held a series of meetings in 1929 and 1930 that culminated in a proposal:[26]

1. Arizona could fully use the waters of the Gila River system and its other tributaries, except for return flows to the main stem.

2. The Lower Basin's 7.5 million acre-feet of III(a) water would be apportioned as follows:

    a. California: 4.4 million acre-feet

    b. Arizona: 2.8 million acre-feet

    c. Nevada: 0.3 million acre-feet

3. The Lower Basin's 1 million acre-feet of III(b) water would be supplied from the main stem and apportioned as follows:

    a. California: 0.5 million acre-feet

    b. Arizona: 0.5 million acre-feet

4. The available surplus (estimated to be 2 million acre-feet) would split fifty-fifty between Arizona and California.

5. California and Arizona would share the Mexican burden (estimated to be 800,000 acre-feet per year in 1930) on a fifty-fifty basis.

6. Shortages would be shared on a proportional basis without regard to priority.

litigation. Arizona likely would have ratified the 1922 compact, avoiding four rounds of Supreme Court litigation in the 1930s, and there might not have been a future dispute over a water supply for the Central Arizona Project, avoiding additional litigation in the 1950s.

Of course, the magic of the proposal was that it was based on the optimistic hydrology used to market Hoover Dam's water and power, significantly overstating the amount of water available on the lower Colorado River for future use.

It was not to be. California rejected Donovan's scheme, asking for more water. It wanted all of the III(b) water, arguing that it was needed to meet their planned water contracts.[27] Arizona rejected that proposal, leaving the process at a stalemate, and leaving Secretary of the Interior Wilbur no choice but to proceed with the completion of water contracts with California parties.

CHAPTER 9

The 1930s—Drought, Development, and Litigation

In the summer of 1930, as Secretary of the Interior Lyman Wilbur sent his one-line telegraph to Commissioner Elwood Mead giving the go-ahead for construction of the world's largest dam, the Colorado River was having another good year. The nation was not.

The U.S. unemployment rate in 1930 jumped from 2 percent to nearly 12 percent. Desperate families converged on southern Nevada in hopes of a job helping build Hoover Dam.[1]

In July 1930, as Wilbur and Mead were launching their ambitious project, Colorado River flows past the Black Canyon dam site were good. The premise on which the great project was based seemed sound. A year later, things looked very different. In July 1931, as Wilbur trekked to the remote dam site to push the button setting off the ceremonial first construction blast, flows were less than half of the year before. In the Colorado River Basin, 1931 still stands as one of the seven driest years on record.[2] It was the beginning of the first great drought of the twentieth century, a decadal shift in the basin's climate that would have a profound and lasting impact on the plans even then still being laid. But of the two forces—the economy and the climate—it was the economy that drove the project, the belief that building Hoover Dam and the economic development that would go with it was of paramount importance. The grandiose project gave the nation hope at a dark time. The

New York Times waxed eloquent about the task ahead, telling the fanciful tale of a hard-rock engineer walking to the edge of the cliff above Black Canyon and looking down at the river he was about to tame. "The fellow who stole fire from heaven was hardly more presumptuous," the *Times* wrote.[3] Any concerns about drought on the Colorado, and the risks that might pose to the long-term viability of the enterprise, were swept aside.

But drought it was, and it was only a matter of time before the presumptuous task of taming the Colorado River would have to take it into account. Just as Secretary Wilbur was completing the power and water contracts with the California agencies that would assure the repayment of Hoover Dam's construction bills, the Colorado River entered a sustained dry period.

The 1930s began well enough. In 1930, the natural flow at Lee Ferry was 14.6 million acre-feet, in line with what we now know to be long-term average. But in 1931, it fell to 8.6 million acre-feet, the lowest flow since gauging stations were installed in the 1890s. In 1932 and 1933, flows were average to slightly below average, but still no cause for alarm. In fact, low flows were welcome to the construction crews at the Hoover Dam site. In 1934, the bottom fell out when the estimated natural flow at Lee Ferry was only 6.63 million acre-feet and the actual flow at the Hoover Dam site was less than 4.5 million acre-feet. In the Colorado River Basin, 1934 was the second driest year in the twentieth century. Over the course of the 1930s, only two of ten years had flows above the 18.5 million acre-feet per year Debler had projected as the critical flow to make the project's economics work.

In 1935, as the Bureau of Reclamation began filling the reservoir behind its new dam, the Colorado River Basin had unequivocally descended into the first extended drought of the twentieth century, a stark hydrologic reality demonstrating that the cautions of LaRue, Stabler, Sibert, and their colleagues had been right.

Even as users were beginning to take the first water from the resulting projects, the drought of the 1930s forced a reckoning with two important implications. The first was that, even as the first of the great dams on the lower river was being finished, more would be needed. In order to comply with the 1922 compact's delivery requirements, 75 million acre-feet every ten years plus the as-yet-undetermined share of the U.S. delivery obligation to Mexico, the Upper Basin would need sufficient storage of its own—as much as 28 million acre-feet of carry over storage—extra water from earlier wet periods, saved for the dry.[4] It meant that more great dams, rivaling Hoover Dam in size and complexity, were in the basin's future.

The second new reality related to the surplus, now forgotten but embedded in and essential to the political bargain. The 1922 compact had only carved up a portion of the river's water, leaving what the commissioners thought would be a surplus to be allocated later. Recall that Chairman Hoover, Colorado's Carpenter, and the other commissioners all believed that the Colorado River system would produce on average 20–22 million acre-feet per year. The compact was, in part, sold under the concept that it only apportioned 16 million acre-feet, therefore another four to six million acre-feet were available as surplus for Mexico and future apportionment.

When the 1930s began, the consensus of the Colorado River water establishment was that the long-term natural flow of the Colorado River at Lee Ferry was at least 17.5 million acre-feet per year. By the end of 1943, the drought of the 1930s had intruded on the record the river's managers were using to estimate its flows. Reclamation's estimate for the long-term natural flow at Lee Ferry had dropped to 16.27 million acre-feet per year, well below the 17.5 million acre-foot 1920s benchmark, but still almost 10 percent higher than the 14.8-million-acre-foot-per-year modern estimate for 1906–2016 and nearly 15 percent higher than Sibert's conservative estimate of 14.2 million acre-feet.

Reclamation had the information available to be more conservative in its approach. The 16.27-million-acre-foot average is based on a period of record that includes three decades of above average flows: the 1900s, the 1910s, and the 1920s, and only one decade of below-average flows—the 1930s. Yet again, the below-average decades of the 1880s and 1890s were ignored. Had Reclamation considered them, the agency would have concluded the long-term mean flow at Lee Ferry was about 15 million acre-feet per year, close to today's figure. The disincentive was clear. Reclamation engineers and planners were now joined to the "iron triangle," working with boosters and politicians to complete the grand plan for full development of the basin. Less water would have meant more conflicts among the boosters and politicians and fewer projects for the Bureau of Reclamation, now basking in the success of Hoover Dam.

Construction Moves Forward

At the bottom of the system, work on the All-American Canal to carry water to California's Imperial Valley began in the summer of 1934, followed shortly by Imperial Dam, the concrete structure spanning the river twenty miles upstream from Yuma that diverts water into the All-American Canal. Combined with Hoover Dam's role in flood protection and water storage, the All-American Canal and Imperial Dam ensured the future of Imperial Valley agriculture.

Equally impressive were the efforts by the Metropolitan Water District of Southern California (MWD) to complete its Colorado River Aqueduct and Parker Dam that went with it. MWD's success meeting the water needs of Southern California's booming population gave birth to an agency that would be both respected and feared by the rest of the basin. With the completion of Hoover Dam, the All-American Canal, Parker Dam, and the Colorado River Aqueduct, by the early 1940s California had in place the projects necessary to use over 5 million acre-feet per year of Colorado River water. Hoover Dam was the keystone. Lake Mead provided storage for flood control and river regulation, and its turbines generated electricity to pump water to the coast. The quick completion of these projects also validated the concerns of Colorado's Delph Carpenter and his fellow Upper Basin commissioners. Without a compact, and with the application of prior appropriation basinwide, these large California projects could have commanded the entire river in at least the drier years and would have been an obstacle to development of Colorado River water on the upper river.

While the projects were not of the size of those in the Lower Basin, the 1930s also brought development to the Upper Basin, particularly in Colorado. The largest project was the Colorado–Big Thompson project (C–BT). The C–BT project was conceived in the early 1930s by irrigators in the South Platte River Basin, to the east and across the continental divide from the Colorado River Basin. By that time, all of the native South Platte water, except for wet years, had been appropriated. When the wet decades of the 1910s and 1920s ended with the drought of the early 1930s, there were major shortages on the South Platte. Thus, the farmers looked across the continental divide to the headwaters of the Colorado River in Grand County.

After internal debate within Colorado between representatives of north-eastern Colorado and the West Slope and a 1937 compromise, the project was authorized by Congress on August 9, 1937. The C–BT project is not a classic Bureau of Reclamation project that provides a water supply to reclaim lands for irrigation purposes. Instead, it was designed to provide supplemental water to lands already under irrigation that had an insufficient local supply. No new lands were irrigated by the project.

The Bureau of Reclamation signed a repayment contract with the Northern Colorado Water Conservancy District in 1938. Construction started shortly thereafter. Due to World War II's interruption and the complexity of the project, the first water deliveries did not occur until 1947, and the project was not fully completed until 1957.[5]

The C–BT project shares a common theme with some of its Lower Basin cousins. Its water yield was significantly overestimated. The project was designed for the delivery of 310,000 acre-feet per year, but after more than sixty years of operation it averaged about 213,000 acre-feet. There are several reasons for this, including modifications to the project and required bypass flows, but most importantly, like the 1922 compact or the Hoover Dam water contracts, the period used by Reclamation to evaluate the project hydrology has now been shown to be unusually wet. After the project was built, the East Slope beneficiaries of the C–BT would even take the benefits of this optimistic hydrologic period of record one step further and consider the 310,000 acre-feet of supply more than a projection. They considered it a legal entitlement.[6]

The C–BT project was the biggest, but not the only, Upper Basin project that would see construction start in the 1930s. In western Colorado, Reclamation built Taylor Park Dam and Reservoir on the Taylor River above the town of Gunnison and Vallecito Dam and Reservoir on the Los Pinos River near Durango. Denver Water began building its Moffat Tunnel to move water from the Upper Colorado watershed across the continental divide to the South Platte. Arkansas River irrigators built the Twin Lakes (also known as Independence Pass) project to divert waters from the headwaters of the Roaring Fork River. Both of these were "nonfederal" projects but received considerable federal financial support through New Deal programs.

The dry hydrology of the 1930s did have an influence, bringing a more realistic view to a number of important policy decisions that would be made in the late 1940s through the late 1960s. In a 1945 letter to Senator Albert Hawkes of New Jersey during the debate over the ratification of a treaty with Mexico, former commission chairman Herbert Hoover said this: "The longer the period of stream-flow records, the less becomes the safe yield of the river in extended low flow periods. . . . There can only be one conclusion: That as time passes, the safe water supply of the Colorado River has found to grow less, while the requirements for, and the value of, that water increase many-fold."[7]

While Other Basin States Begin Building, Arizona Goes to Court

While California and Colorado charged ahead in the 1930s, launching projects to bring Colorado River water to their farms and cities, Arizona spent the decade trying to build its legal case that it had been wronged by the deals cut in the 1920s. Arizona's efforts came to naught, but they nevertheless represent a critical episode in the evolution of Colorado River management, as Arizona struggled to make sense of the complex relationships between the institutions being built and the reality of the river's hydrology.

Arizona's efforts, laid out in lawsuits filed with the U.S. Supreme Court in 1930, 1934, and 1935, show the state's water leaders wrestling with confusion over the state's use of water from the Gila River, and how that water should be accounted for under the 1922 compact and the 1928 Boulder Canyon Project Act. All three of the suits were unsuccessful, but the evolution of Arizona's arguments about the relationship between Colorado River hydrology and law still echo in river management difficulties today.

Initially, in 1930, Arizona attacked the constitutionality of the 1922 compact and the 1928 act on the grounds that they fail the reasonable test of "equitable apportionment." Arizona argued that the 1922 compact gave the Upper Basin more water than it could use and the Lower Basin less water than it needed and the 1928 act gave all of the remaining Lower Basin water to California.[8] The Supreme Court moved swiftly, rejecting the argument just seven months after the case was filed, but Arizona was undeterred. It shifted ground in its next two lawsuits, arguing for a new definition of water "use" that would greatly increase the amount of Arizona's actual water use that could be charged against its Colorado River Compact allocation.

The heart of Arizona's 1934 and 1935 claims involved the Gila River and its tributaries, the Salt and Verde, and the compact's Article III(b), which set aside the additional million acre-feet of water for Lower Basin use. The authors of the 1922 compact and the 1928 act never fully came to terms with the fate of the Gila's water, leaving an ambiguity in the rules that opened the door to decades of litigation, on which hung the fate of huge water development projects in both Arizona and California.

The Gila, Salt, and Verde flow from the high county of central Arizona and southwest New Mexico into Arizona's central valleys, delivering what Arizona estimated at the time was 3 million acre-feet of usable water. By the mid-1930s, farmers were using that water to irrigate 525,000 acres of Arizona farmland.[9] In its initial legal assault, Arizona argued that the language of the Colorado River Compact included all of that water in the definition of the "Colorado River System" ("that portion of the Colorado River and its tributaries within the United States of America"), effectively depriving Arizona of any other Colorado River water.

If Arizona used water in Phoenix that never would have made it to the main stem of the Colorado River anyway, should that be counted against Arizona's share of Colorado River "use"? If Arizona could claim this "use" did not count against its Colorado River allocation, it would be able to develop far more water for its farms and cities, to the detriment of supply available for California.

In a table submitted in its 1935 suit, Arizona implicitly offered a new way of accounting for its use of water from the Gila River and its tributaries. The full 3 million acre-feet of Gila use it noted in its 1930 filing should not be the real measure of Arizona's water use; rather, it only should be charged the 1.3 million acre-feet of natural flow that made it all the way down through the deserts to the junction of the Gila and the Colorado, just upstream from Yuma.

The Supreme Court was unmoved, again denying Arizona's claims, but in floating the first tentative version of this new version of Colorado River water accounting, Arizona was laying the groundwork for a struggle over how to interpret the relationship between hydrology and law that would persist for decades.

Thus, by the middle of 1936, Arizona was still opposed to the ratification of the 1922 compact but had made no progress in its attempts to upend it. It had challenged the constitutionality of both the 1922 compact and 1928 act and failed. Hoover Dam was nearing completion. Parker Dam, the Colorado River Aqueduct, and the All-American Canal were under construction, and the Supreme Court had refused to give Arizona an equitable apportionment of the Colorado River. Arizona had no realistic choice other than to ratify the compact. In the early 1940s, Arizona would turn its attention to obtaining a satisfactory contract with the secretary of the interior for Colorado River water, but its conflicts with California over the Colorado River remain largely unresolved, even today.

In March 1943, the Arizona legislature ratified the compact contingent upon a satisfactory contract between the secretary of the interior and the State of Arizona for the delivery of 2.8 million acre-feet of water from Lake Mead. Over the objections of the State of California, the contract was signed on February 9, 1944. The contract was carefully written so as to avoid any controversial issue regarding the amount of water available to Arizona, or to any compact state, under the 1922 compact and the 1928 act.

The contract provides Arizona with "so much water as may be necessary for the beneficial consumptive use for irrigation and domestic uses in Arizona of a maximum of 2.8 million acre-feet" per year.[10] In addition to the 2.8 million acre-feet, the contract allows the secretary to deliver one-half of surplus waters "unapportioned" by the 1922 compact and available for use in Arizona. Thus, as with the California contracts, if sufficient water was available, the United States accepted the channel and delivery losses. Under the contract, Arizona recognizes the California contracts with the secretary of the interior and the rights of Nevada, Utah, and New Mexico to utilize Lower Basin water under the 1922 compact. The entire contract was subject to reservations providing that nothing in the contract was intended to interpret or prejudice any of the contested provisions of the 1922 compact or 1928 act.[11]

Upon completion of the contract, the Arizona legislature unconditionally ratified the 1922 compact. The legislation was signed by Governor Osborn on February 24, 1944. The stage was now set for the next phase of Colorado River development—coming to an agreement on Mexico's share of the river and finding a way for Arizona to get its share.

CHAPTER 10

The Treaty with Mexico

Another dry year on the Colorado River, 1940, was the trough of a drought that pushed the ten-year average flow on the Colorado River to 12.5 million acre-feet per year.[1] The drought of the 1930s should have been a reminder of the problems left by the failure to listen to LaRue, Stabler, and Sibert when they questioned the assumptions on which the edifice of Colorado River development had been built. Instead, as the boosters headed into the next phase of Colorado River development—the crucial decision of how much of the shrinking river would be allocated by treaty to Mexico—the river's federal managers, with the acquiescence of the states, followed a familiar path. In order to avoid showing the Senate a declining trend in the flow of the Colorado River that could have raised significant opposition to the treaty, they fudged the numbers.

But in the debate over the Mexican treaty that followed, a new political dynamic began emerging that would persist through much of the second half of the twentieth century as paper water estimates increasingly conflicted with hydrologic reality. California, with the dams and canals in place to use its share of the river's water, began to flex its formidable political muscle to try to block Mexico and the U.S. states from projects that would eat into the water supply already flowing to the Imperial Valley and coastal Southern California. Whatever the inflated paper estimates of supply may have suggested, California was behaving as if the Colorado River's supply was limited after all.

In the fall of 1940, Mexico's ambassador sent a letter to the U.S. State Department about a matter that was of increasing urgency to the nation sitting at the southern terminus of the Colorado River. Francisco Castillo Nájera's concern had a familiar ring. Much as Arizona and the Upper Basin states had looked on with alarm as California's use of Colorado River water grew, Mexico felt increasingly threatened by upstream development on the river, which could leave it without the water it felt it needed and to which it believed it was entitled. Development of its agricultural lands depended on a reliable supply of water, and a failure to settle the question of how much Colorado River water Mexico could depend upon was therefore "an economic aspect of great importance," Nájera wrote in a November 16, 1940, letter. "The passage of time works unfavorably for my country and the interests being created in the United States continually make an adequate solution more difficult."[2]

By the 1922 Colorado River Compact's Article III(c), the negotiators had acknowledged the likelihood that water would be allocated to Mexico. They structured the compact to provide a future "surplus," believing there was sufficient water for the 1922 compact's apportionments to each basin and for that surplus to satisfy both Mexico and future additional apportionments within the United States. The debate by the 1940s revolved around the question of how much "surplus" might be available without cutting into the shares of the seven U.S. basin states. Any attempt at a U.S.-Mexico treaty, negotiated at the federal level but then subject to approval by the United States Senate, would need support within the basin to overcome that obstacle. The states had to be convinced that there was still enough water in the river to pull off such a deal.

The specter of a water treaty with Mexico had been looming over the basin since the late 1920s. The issues were bigger than just the Colorado River. Mexico and the United States also share the Rio Grande. Both Texas and Mexico had a large amount of irrigated agriculture on the Lower Rio Grande, and since most of the water originated in Mexico, the irrigators in Texas needed Mexican cooperation. The connection between the Rio Grande and the Colorado was a geopolitical reality that water managers in the Colorado River Basin resented but could not change. Commissioner Elwood Mead chaired a three-man United States negotiating team that made an unsuccessful treaty effort in the late 1920s.

A decade later with a global war expanding and U.S. relationships with Mexico and the rest of Latin American of crucial strategic importance, the U.S. State Department had motives entirely unrelated to the issues of water management.[3] So the federal government and its basin state allies again did what they had done before when the risk of an inadequate Colorado River supply threatened its larger agenda—misused the available data concerning the long-term flow of the Colorado River, fudging the numbers to make it look like the surplus was still large enough to satisfy Mexico and the full apportionments to each basin under the 1922 compact. Unfortunately, depending on how apportionments were measured, the water to Mexico could seriously cut into California's water supply.

By the time the negotiations were completed four years later, the threat of world war had been replaced by the diplomatic needs of a postwar world and the battle against communism, but the incentives remained the same. A shortage of water in the Colorado River would not be allowed to stand in the way of national goals.

By the time of Nájera's 1940 letter, discussions between the two nations had been dragging on fitfully for the better part of a decade. As they lingered, the reality of the dry 1930s and the contracts between the federal government and California agencies for 5.36 million acre-feet per year of water made it harder to strike a bargain. As a basis for negotiations, Secretary of State Cordell Hull requested a report to help determine how much "surplus" Colorado River water might be available for Mexico. The confidential report by consulting engineers Joseph Jacobs and John Stevens, completed in 1938, became another milestone in government efforts to nail down the question of how much water the Colorado River really had to offer. Importantly, this was the first report since the work of LaRue, Stabler, and Sibert in the 1920s done by anyone outside the Bureau of Reclamation.[4]

The report provides a good summary of how the Bureau of Reclamation and the basin states viewed the supply and demands for Colorado River water as of 1938. Jacobs and Stevens used a figure of 17.85 million acre-feet for the natural flow of the Colorado River at the border with Mexico.[5] This

Table 4 Declining estimates of the Colorado River's natural flow at Yuma

Source	Estimated flow
Carpenter's December 15, 1922, report to the Colorado legislature	20.5 million acre-feet
Debler's 1930 report, used for the Hoover Dam contracts	20 million acre-feet
Arizona's 1935 "equitable apportionment" case claims	18.2 million acre-feet
Jacobs-Stevens 1938 report	17.8 million acre-feet

reflected the river through 1937 and continued the downward decline of the estimated natural flow of the Colorado River at the border. But again, all four estimates ignored the droughts of the late 1800s identified and quantified by Stabler and LaRue.

Depending on development in the basin, Jacobs and Stevens estimated a surplus of between 1.3 million and 1.6 million acre-feet of water available for Mexico. Much of the water would come from return flows from irrigation in the Yuma area and from river flow that would be used to flush the silt away from the All-American Canal diversion silting basins.

Copies were provided to members of the Committee of Fourteen, two representatives from each of the seven Colorado River Basin states. This group was formed to advise and provide input to the federal government on the treaty negotiations and was playing a crucial role in basin water management at the time.

The Committee of Fourteen was highly critical of the report. There was too much uncertainty to reliably forecast how much water could be dependably provided to Mexico, they concluded, asking that any action on a treaty be delayed until studies then underway on the expansion of basin irrigation projects were completed.

As consulting engineers with no real stake in the politics of the basin, Jacobs and Stevens put the committee in an interesting position. They brought in the perspective of outside experts. As such, the report became one of the more realistic pictures of the basin's future, but one that was difficult to reconcile with the political aspirations of the individual basins and states. They concluded that the likely development of the basin would mean that the Lower Basin would use more than its annual 8.5-million-acre-foot apportionment, 9.1 million, and the Upper Basin less, only 6 million acre-feet. Interestingly, the committee neither disputed nor discussed the implications of this conclusion.[6] The committee didn't argue with the amount of return flow, but made the case that the water could be made available for use in the United States by collecting it and pumping it uphill to new lands within the United States, rather than sending it downstream to Mexico. The committee also raised something that would in later years become one of the biggest challenges to keeping the Colorado River Basin's water accounts in balance: "Transmountain diversion projects which would have been considered infeasible 20 years ago have been approved and financed, and in some cases are now under construction."[7]

But once again, the committee failed to question the basic assumptions about how much water the Colorado River had. The group made no comment concerning Jacobs and Stevens's conclusion that the natural or virgin flow of the Colorado River at the boundary with Mexico would be 17.85 million acre-feet per year, despite their cautions about the "assumptions to the dependability of relatively short stream flow records as an index to future run-off."[8]

Negotiations Begin

In response to the Nájera letter, the State Department initiated formal negotiations.[9] A Committee of Fourteen meeting was called for May 6, 1941. At this meeting, despite continuing concerns that the comprehensive study had not been completed and a request by Arizona to delay negotiations until the issues between Arizona and California were settled, the committee voted unanimously to approve negotiations with Mexico. California insisted that Mexico be limited to the amount of water it was using as of 1936, about 750,000 acre-feet per year, and the other states agreed.[10]

The State Department negotiators understood that for an agreement to be ratified by the Senate it would need the support of at least most of the basin states. The dynamics of the negotiations within the United States were not simple. The issue was complicated by the fact that the negotiators were trying to settle issues on not one river but three. In addition to the trade-offs over the Colorado River, the issue had been bundled with outstanding international questions on two other rivers, the Rio Grande and the Tijuana. While the Tijuana was straightforward, the Rio Grande also posed major challenges. Further, on the Colorado alone there were major differences among the Colorado River Basin states. Arizona had yet to ratify the 1922 compact. Arizona and California had major disputes that were not resolved and would not be for twenty years or more. Indeed, many of those disputes have yet to be resolved today.[11] All of the basin states fully understood that California, since it had signed contracts for 962,000 acre-feet more than its basic 4.4-million-acre-foot apportionment under the 1928 act, had the most to lose. It was the Upper Basin states that most wanted a settlement with Mexico, fearing that without a treaty Mexico might ultimately acquire more water through international arbitration and possibly impact the amount of water available to the Upper Basin under the 1922 compact.[12]

In July 1941, the United States submitted a treaty proposal to Mexico that provided for 900,000 acre-feet per year of Colorado River water. Mexico rejected the U.S. offer, countering with a request for 2 million acre-feet per year. At a June 1942 meeting of the Committee of Fourteen, Royce Tipton, a consulting engineer who at the time was working for both the State of Colorado and the State Department and would become one of the most important analysts of the river in the years to come, presented a proposal to the states where Mexico would receive 1.5 million acre-feet per year, 900,000 acre-feet of river deliveries plus 600,000 acre-feet of return flows.[13] Led by opposition from California and Utah, the states were unpersuaded by Tipton's argument. The committee ended up suggesting a proposal that would provide Mexico with the 600,000 acre-feet of return flows plus 800,000 acre-feet of deliveries, but the 800,000 acre-feet would be adjusted up or down by 15 percent of the difference between actual deliveries from Lake Mead and 10 million acre-feet.[14]

The State Department refused to take the Committee of Fourteen's alternative back to Mexico as a formal offer but continued to get input from Tipton and Colorado's Clifford Stone. By 1943, the negotiations were settling on a delivery of 1.5 million acre-feet per year, which in turn was cementing

California's opposition. Meanwhile, the Bureau of Reclamation was finishing its own comprehensive study of basin water development, and the Upper Basin states were looking beyond Mexico to an Upper Basin compact and a postwar water development strategy. Arizona, which for hydrologic reasons could have easily been an ally of California in opposing a treaty with Mexico, was holding its cards close. It was making progress on a Colorado River contract with the Department of the Interior, and it too was looking past Mexico to a future where it would have to ratify the 1922 compact and seriously engage California in a battle over the use of the Colorado River.

The Committee of Fourteen addressed many issues beyond a treaty with Mexico, but once again it never questioned the adequacy of the Colorado River's long-term reliable flow. No one brought up the questions raised by LaRue, Stabler, and Sibert about the droughts of the 1800s. With the U.S. Geological Survey out of the picture and the U.S. Bureau of Reclamation left as the only authoritative source of technical information on the river's flow, the common understanding of the 1940s, that the Colorado River had a flow of about 18 million acre-feet per year at the border, was never questioned.

By late 1943, the negotiators for the United States and Mexico had settled on a proposed treaty. Mexico would get 1.5 million acre-feet per year in normal years, a number very close to the recommendation of the Jacobs-Stevens report and the amount suggested by Royce Tipton. The treaty provides that the water could be delivered from "any and all sources."[15] The treaty also provides for the delivery of 1.7 million acre-feet per year when there is a surplus and has a shortage provision that requires Mexico to share shortages on a proportional basis with U.S. users during "extraordinary droughts" or irrigation system emergencies. The treaty does not define *surplus* or *extraordinary drought*.

The Committee of Fourteen met one final time in January 1944 to discuss the treaty. All the basin states except California commended the State Department for obtaining a better treaty than they had anticipated.[16] President Franklin Delano Roosevelt signed the treaty on February 3, 1944. Within weeks, Arizona both ratified the 1922 compact and signed a Hoover Dam water supply contract for 2.8 million acre-feet per year with the secretary of the interior. The stage was now set for the next round of the battle, ratification of the treaty by the United States Senate.

The Ratification Battle and a Basin States Realignment

Arizona's ratification of the 1922 compact and the signing of the treaty with Mexico changed the political alignment within the Colorado River Basin. Led by Colorado, the Upper Basin states along with Texas were strong supporters of the treaty. California adamantly opposed it. Arizona and Nevada were the swing states. Nevada ultimately joined California and opposed the treaty, although its opposition was nominal. Arizona, sharing surplus water with California under the 1928 act, could have easily been an ally with its neighbor to the west, but it was not interested in the surplus. Its primary interests were protecting its development on the Gila River and securing a main stem water supply for a project that could move water from Lake Havasu inland to central Arizona. California, not Mexico, was its primary obstacle. An alliance with the Upper Basin made sense. In January 1945, Royce Tipton, working on behalf of the state of Colorado, wrote a memo laying out the technical foundations of an alliance. He suggested the Upper Basin states and Arizona agree on the definition of how the basin would measure 1922 compact apportionments—a method of accounting for the use of the river's water that would benefit the Upper Basin and Arizona at California's expense.[17]

Senate Hearings

Senate hearings did not begin until January 1945, almost a year after the treaty was signed. Six of the eight states with direct interests supported the deal. Two—California and Nevada—were opposed. The hearings lasted nearly a month. They were chaired by Texas senator Thomas Connally, a strong supporter.[18] Many technical issues, such as the amount of return flows and the actual uses in Mexico were debated, but again there was little serious discussion of treaty supporters' basic assumption that about 18 million acre-feet of water was available for use from the Colorado River system.

At the beginning of the hearing, L. M. Lawson, the American commissioner for the International Boundary Commission (IBC) and one of the

Table 5 Colorado River natural flows presented to the Senate

Year	Period	Source	Annual average flow
1922	1903–20	Senate Document 142	18.11 million acre-feet/year
1929	1895–1922	Senate Document 186	18.38 million acre-feet/year
1934	1897–1922	USBR report	18.171 million acre-feet/year
1937	1902–37	Jacobs-Stevens report	17.85 million acre-feet/year
1944	1897–1943	USBR and IBC	18.131 million acre-feet/year

Source: Statement of Lawrence M. Lawson, Hearings on Treaty with Mexico Relating to the Utilization of Waters of Certain Rivers, Before the Committee on Foreign Relations, Seventy-Ninth Congress, 1945.

lead U.S. negotiators, presented an overview of the Colorado River Basin that included information on the water supply. He presented a chart titled "estimates of virgin flow of the Colorado River (at Yuma) for various periods."[19]

In the long history of the estimates that in one way or another overstated the flow of the Colorado River, Lawson's testimony is the closest one can find to outright fraud. Lawson's testimony obscured the 1930s drought and the resulting decline in the natural flow at Yuma. By mixing actual and natural flow estimates, Lawson was able to conjure enough water on paper to show no downward trend in the estimated annual natural flow in the river between 1922, when the compact was negotiated, and 1944, when the treaty was negotiated, as if the drought of the 1930s had never happened.[20]

Had the table been correct, the Senate committee would have seen that the estimated natural flow of the Colorado River at Yuma continually decreased over a two-decade period from 20.5 million acre-feet per year to 17.8 million acre-feet per year, a 14 percent reduction.

California put on an impressive set of witnesses to testify against the treaty, including Governor Earl Warren, who would later become chief justice of the U.S. Supreme Court; Attorney General Robert Kenny, former congressman and House sponsor of the 1928 act; Phillip Swing; and a host of managers, attorneys, and board members from its agencies that used Colorado River water. None of these witnesses would seriously challenge the underlying hydrology as presented by the treaty's supporters.

There were some skeptics. Former President Herbert Hoover and George E. P. Smith, still a professor of irrigation engineering at the University of

Arizona, questioned the hydrology. In a last-minute attempt to derail the treaty, California called on Hoover to oppose it. He did so in a March 17, 1945, letter to Senator Albert Wahl Hawkes of New Jersey arguing that the long-term flow of the river had been declining from the levels he and his fellow compact commissioners had before them in 1922.[21]

Smith opposed the treaty, criticizing the hydrologic assumptions used to justify the treaty. He believed the long-term average natural flow at Lee Ferry was about 14.5 million acre-feet per year, almost 2 million acre-feet less than the Bureau of Reclamation 1944 estimate of 16.27 million acre-feet. He also believed treaty supporters were intentionally overstating the actual amount of lands in Mexico that were then under irrigation. However, he did not publish his hydrologic criticism until more than ten years later.[22]

The Senate approved the treaty on April 18, 1945, six days after the death of President Roosevelt, by a vote of 76–10. From the seven Colorado River Basin states and Texas, the Senate vote was 12–4, with California and Nevada senators voting no. Mexico did not begin its ratification process until after the United States Senate had voted. In fact, the treaty was not even released to the public until two days before the United States Senate final vote. After its own internal debate, Mexico's Senate ratified the treaty on September 27, 1945. The treaty entered into force on November 8, 1945.[23]

CHAPTER 11

The Argumentative Wisdom
of George E. P. Smith

George E. P. Smith is another expert who got the hydrology right, for the right reasons. He believed the data being used to support Colorado River development was based on an unrepresentative period of time and flawed measurements. Like LaRue, Stabler, and Sibert before, his message was inconvenient, so he was largely ignored.

Smith (1873–1975) was a longtime professor of irrigation engineering at the University of Arizona. He was born in Vermont and educated at the University of Vermont before coming to the University of Arizona in 1900, twelve years before Arizona became a state. He was a professor of engineering and physics for six years, then a professor of agricultural engineering until his retirement in 1955. He was considered an authority on irrigation and a pioneer in the development of the underground and surface waters of the state of Arizona. He wrote the first Arizona water code, which was enacted by the Arizona legislature in 1919.[1]

His critique of the treaty with Mexico, *Arizona Loses a Water Supply*, published in 1956, is largely a collection of his letters to Arizona officials concerning the treaty between the United States and Mexico.[2] He was a major critic of Arizona's support for the treaty. He believed that the 1944 estimated natural flow of 16.27 million acre-feet per year at Lee Ferry was as much as 2 million acre-feet too high. He had no confidence in the gauge records and stream flow reconstructions before 1922. Therefore, he recommended using a period of record of 1922–43 with an average natural flow of 14.5 million acre-feet per year for determining the Colorado River's water supply.[3] Although he recommended a figure based on a short period of record, his 14.5-million-acre-foot recommendation is in line with today's long-term estimates.

Smith had a Lower Basin and an Arizona perspective. He considered it a mistake for both California and Arizona to agree to the 1922 compact. He believed they should have been allies in water development, not enemies. He showed considerable disdain for Delph Carpenter. He referred to him as "Mr. Troublemaker" and concluded "it is true that California in submitting to the Delph Carpenter blackmail in 1922 dragged Arizona into a helpless and hopeless position. But since the early thirties the interests of the two states have been parallel and consisted in inducing or forcing the upper basin to relinquish some of its unneeded allocation of water." He believed that the Upper Basin would ultimately consume only 4.5 million acre-feet per year and the Lower Basin would end up consuming over 11 million acre-feet per year, over 2.5 million acre-feet more than its 8.5-million-acre-foot apportionment under the 1922 compact. Those figures are very close to today's situation on the river.[4]

Like his contemporary E. C. LaRue, Smith's personality and communication style limited his effectiveness. He was brash, argumentative, had a high opinion of his engineering skills, and appeared to have had little regard for the lawyers and politicians engaged in the water issues of the time. Smith was particularly incensed with the claim by the supporters of the treaty that Mexico was, as of 1944, consuming 1.8 million acre-feet of water per year to irrigate its lands. Supporters argued the treaty, which provided Mexico with 1.5 million acre-feet in normal years, benefited the United States by reducing Mexico's use of Colorado River water. Smith believed that actual uses in Mexico were in the 750,000-to-900,000-acre-foot range.[5] He was especially critical of Charles Carson, attorney for the Colorado River Commission of Arizona. Carson was one of the leading witnesses for the six states that supported the treaty. In his book, he states, "In my letters to the State Land and Water Commissioner, as shown above, I dissected, deflated and demolished Mr. Carson's report on water supply. Despite this, the state officials and Arizona's two U.S. senators clung to the Carson report and even brought Mr. Carson to Washington to testify." In a footnote he adds, "The scant consideration given to my letters has puzzled me. It was the report of a recognized hydrologist against the report on hydrology made by a lawyer."[6]

Smith was critical of the Bureau of Reclamation's estimates of the river's flow before 1922. In reference to a 1953 supplemental report on the hydrology of the Colorado River, he writes, "The presentation of the data in those tables, especially in the Supplement, makes no distinction between the years when flow was actually measured and the earlier years. The general reader would infer that all records back to the earliest ones were measured flow. That is intellectual dishonesty. . . . Stream gauging in Arizona prior to 1922 was extremely sketchy. . . . When the unreliable discharge data preceding 1922 were used to synthesize a record for Lee Ferry the errors were compounded. Why the Bureau does not discard the synthetic records has no explanation."[7] Interestingly, Royce Tipton, who was the primary engineering witness for the treaty supporting states, would come to the same conclusion in the mid-1960s when Congress was considering the Central Arizona Project.

Like LaRue, on a number of critical issues, Smith was wrong. The theme of Smith's 1956 report is that Arizona officials had a long history of mis-managing the states' Colorado River resources and, therefore, there was no water supply for the Central Arizona Project. He concludes that if Arizona is limited to 2.8 million acre-feet of main stem water, only 215,000 acre-feet of water is available for export to central Arizona. Therefore, the Central Arizona Project is not feasible. What he got wrong was that he assumed Arizona's Gila Project would be much larger than as ultimately constructed, consuming about 1.2 million acre-feet more than as built. Smith wrongly predicted that Arizona would lose its 1953 Supreme Court case against California. The downsizing of the Gila Project and Arizona's Supreme Court victory made the Central Arizona Project feasible.[8] Looking at his 1956 publication in total, on water supply issues Smith was right more often than he was wrong.

CHAPTER 12

"A Natural Menace Becomes a National Resource"

With the Mexican treaty behind them and the war over, members of the Colorado River Basin community turned in the mid-1940s to the grand plan of full development of the Colorado River's waters. But the effort remained mired in a mix of hubris and frustration.

The hubris came from the success of prewar river development. Hoover Dam had shown the value and popular success of water development on the river. When dignitaries from across the basin gathered in 1946 to celebrate the tenth anniversary of power being generated by Hoover Dam's turbines, a squadron of navy planes flew in from the coast to celebrate the role of the dam's electricity in the war effort. "And although (the) celebration is listed as a sort of birthday gathering, many people are hopeful of other more far-reaching results," the *Los Angeles Times* reported. "They say the gathering at the big dam, where the dignitaries of the States will participate in a common gesture of good will, may well lay the foundation for peaceful settlement of many perplexing problems still attendant upon future equal distribution of Colorado River water."[1]

The unrealized promise was tantalizing. As much as 8 million acre-feet per year of water still flowed unused—"wasted"—past Yuma and into Mexico and the sea. The pressure to develop that water with the economic benefits it would provide, in both Arizona and the Upper Basin, was strong.

Communities saw the benefits California had already reaped, and there was growing acceptance of the idea of federal subsidies for rural development. And rural western states had increasingly consolidated their congressional power on the key committees needed to authorize and finance big Bureau of Reclamation dams. The idea of expanding the web of Colorado River water development projects "enjoyed overwhelming public support, not just among western farmers, but among their city brethren, too; conservatives, liberals, Democrats, Republicans—ideology meant nothing where water was concerned," wrote Marc Reisner.[2]

But there still were, as the Los Angeles Times had written, "perplexing problems" standing in that development's way—the states' persistent inability to sort out the question of allocation. Who was entitled to how much water? The 1922 compact and the 1928 Boulder Canyon Project Act had provided the starting allocation framework, but the states remained unable to complete the next steps needed to sort out the remaining questions. Until they did, the "waste" would continue.

California remained the beneficiary of that "waste." With its diversions to the Imperial Valley and coastal Southern California completed, failure to develop the rest of the basin's water projects worked to California's advantage. It ensured a surplus of unused water to fill the Colorado River Aqueduct and the All-American Canal. California had an incentive to block future deals, and the political power to act. The flip side of that reality was that the other states had a continued incentive to ignore past droughts and overestimate the river's supply so that their projects could be built. The incentives to continue ignoring LaRue, Stabler, and Sibert were strong.

The Bureau of Reclamation and the Department of the Interior provided the prod to this next phase of development discussions in a 1946 draft report grandly titled The Colorado River: "A Natural Menace Becomes a National Resource"—A Comprehensive Report on the Development of the Water Resources of the Colorado River Basin for Irrigation, Power Production, and Other Beneficial Uses in Arizona, California, Colorado, Nevada, New Mexico, Utah, and Wyoming. It itemized 134 projects large and small to dam and deliver water to more than a million additional acres of agricultural land and generate nearly twenty billion kilowatt-hours per year of electricity across the basin.

Published in the form of a hefty book with tables, maps, and pictures, the report that came to be known as House Document 419 (HD 419) was part technical report and part glossy marketing for an optimistic future. The politically savvy Reclamation commissioner Michael Strauss had packaged a suite of goodies tailored to bring maximum support from members of Congress eager for money to be spent in their districts. The report represented the full flourishing of the "iron triangle" of river development, especially the legs of the triangle represented by Congress and the Bureau of Reclamation. With optimistic assumptions about the economic benefits of irrigation ad power sales, the report concluded the projects taken as a whole "could return to the Nation $1.30 for each dollar required to construct, maintain and operate the projects."[3]

But it came with a catch. None of the projects could even begin, Strauss told Congress in the forward to the report's final version, until the states got their water rights allocation house in order. Between the 1946 draft and the final report's issuance in 1947, even the name had changed. Now rather than a plan for the future, it was merely an "interim report on the status of investigations" into that future.

"Further development of the water resources of the Colorado River Basin, particularly large-scale development, is seriously handicapped, if not barred, by the lack of a determination of the rights of the individual States to utilize the waters of the Colorado River system," Reclamation's Strauss wrote in the report's cover letter. For the projects to go forward, Reclamation had concluded, the agency needed assurance that the supplies of water were sufficient to justify the projects' costs. Reclamation was pressing for compacts among the states within each basin, giving the states certainty about how much water each could use, or lacking that, court action accomplishing the same.[4]

The 1922 compact had specified overall Upper and Lower Basin allocations, leaving the hard task of divvying up each basin's apportionment among the states for later. The 1928 Boulder Canyon Project Act had advanced the issue by suggesting an allocation among Arizona, California, and Nevada, while leaving the difficult task of negotiating an actual implementing compact to the three states. By 1946, Wyoming, Colorado, Utah, and New Mexico were well on their way to a compact of their own, dividing up the Upper

Basin's share.[5] But the Lower Basin states were stuck, mired in a conflict between Arizona and California.

California and Arizona each had plans for over 5 million acre-feet of Colorado River water, far more than the 1922 compact actually allocated to the Lower Basin. More importantly, it was far more than the river could provide. But rather than confront hydrologic reality, House Document 419 shifted the debate to an arcane water accounting argument. Arizona and the Upper Basin states favored an accounting scheme that they believed would allow far more use within their states. Arizona's approach would allow its full development of its 5 million acre-feet of projects, including a "Central Arizona Project" to bring water to the Phoenix and Tucson valleys, and a second large project to bring additional irrigation water to the Yuma area in the state's southwestern corner. But in so doing, Arizona's accounting scheme would eat up the surplus in the river, on which nearly 1 million acre-feet of California's 5.36 million acre-feet of Hoover Dam contracts relied. It would threaten over half of the water supply of Metropolitan Water District's already-completed Colorado River Aqueduct.

The future of the Colorado River Basin was mired in an argument over accounting and the concept of "salvage by use."

The Quirky Math of Salvage by Use

In their formal response to the 1946 draft of House Document 419, the states focused on questions left unsettled by the 1922 compact and the Boulder Canyon Project Act and complicated by the Mexican treaty. How were apportionments to be measured under the 1922 compact? How much surplus was available under the 1922 compact with the Mexican treaty now in force? The common denominator for these questions was the murky concept of "salvage by use."

The concept of salvage by use is real. It is made possible by the geography and climate of the Colorado River system. Virtually all of the river's normal runoff is generated from its higher elevation watersheds. This is where it rains and snows. But once the main stem and its most important Lower Basin tributary, the Gila River, leave the higher elevations, they run through some of the country's hottest and driest deserts. As they make their way through the desert to the Gulf of Mexico, they pick up little additional water from tributaries and lose a lot due to the hot desert sun and vegetation.

The Gila is the best example of salvage by use. The long-term average natural flow of the Gila River system is about 2 million acre-feet per year as it leaves the high country north and east of Phoenix. Allowed to flow naturally to its confluence with the Colorado River at Yuma, it would lose, on average, about 1 million acre-feet per year to riparian vegetation and evaporation. The Gila's contribution to the average natural flow of the Colorado below Yuma thus was only about 1 million acre-feet per year prior to development. By fully using the water upstream near Phoenix, the water that would otherwise be lost to evaporation between Phoenix and the confluence with the Colorado could be "salvaged" and put to beneficial use. Arizona could use 2 million acre-feet of water, but at a cost of only 1 million acre-feet of water to the main stem of the Colorado. The Gila River is not the only example. After the Colorado River left the Grand Canyon, it lost at least 800,000 acre-feet per year as it traveled south to Yuma, and about 400,000 acre-feet could be salvaged.[6] In the Upper Basin, engineering advisors to the Upper Colorado River Basin Compact Commission believed that the Colorado River and its principle tributaries lost about 600,000 acre-feet per year as they flowed from the state lines to Lee Ferry.

Accepting that salvage by use is real, why does it matter? On a river where the states were competing for every drop of water, Arizona and the Upper Basin states believed they could use the concept to game the system and consume more water under their fixed apportionments. Remember that the Colorado River Compact apportions "beneficial consumptive use." The Lower Basin gets 8.5 million acre-feet per year and the Upper Basin 7.5 million acre-feet per year. Section 4 of the 1928 Boulder Canyon Project Act also makes allocations of beneficial consumptive use to the Lower Basin states.[7] Thus, if a basin or state is limited by a fixed amount of beneficial consumptive use, it is important to know how to measure beneficial consumptive use. The most common method used is diversions minus return flows. You measure the amount of water diverted for use for irrigation or municipalities and measure (or estimate) the amount that returns to the stream through groundwater recharge that eventually seeps back into the river, or wastewater treatment plant outfalls, or drain systems at the downstream end of farm systems. That return flow is then available for others. The returned amount, subtracted from the amount originally diverted, is the beneficial consumptive use.[8]

That sounds simple enough, but sometime in the early 1930s after it lost its challenge to the constitutionality of the 1922 compact in the Supreme Court, Arizona figured out a definition that would be to its advantage, and it involved the Gila.[9] Arizona reasoned that if it was using 2 million acre-feet of Gila River upstream, but if the impact of that use to the natural flow of the river at Yuma was only 1 million acre-feet, they could use more water by defining beneficial consumptive use (as it applies to the apportionments under the 1922 compact and 1928 act) as the net impact of all of its upstream uses on the natural flow of the Colorado River at the international boundary. It referred to this method as the "stream depletion" theory. Two million acre-feet of upstream beneficial consumptive would only count as 1 million acre-feet of compact apportionment.

With no Colorado River tributaries, California had reason to disagree. There are limited opportunities in California to benefit from salvage by use. California's largest user, the Imperial Irrigation District, diverts at the bottom of the system near Yuma, but its agricultural tail water flows into the Salton Sea, unusable as return flow. Its other two big users, Palo Verde and MWD's Colorado River Aqueduct, are also far downstream of Hoover Dam, so most of the losses have already occurred by the time the California diverters pick up the water.[10]

Further, because of California's reliance on the surplus, the Arizona definition actually took water away from California. If Arizona was allowed to consume 2 million acre-feet of Gila water within its boundaries, but only have it count against its compact allocation as 1 million acre-feet, that would leave 1 million acre-feet less in the river available for use by others. And the "other" in this case was California and its share of the river's surplus.

In its comments to HD 419, Arizona laid out its argument based on the 1922 compact and the 1928 act. It reasoned that Articles III(a) and (b) apportioned 8.5 million acre-feet in perpetuity to the Lower Basin. Since the 1928 act and the California Limitation Act limited California to 4.4 million acre-feet per year, that left 4.1 million acre-feet of apportioned water for the remaining four states. Subtracting what was owed to Nevada, Utah, and New Mexico, the remainder, 3,669,000 acre-feet per year, is available for Arizona.[11]

Using its stream depletion theory based on the impact to the natural flow of the river at the international boundary, Arizona claimed its depletions were only 1,407,000 acre-feet per year, leaving 2,262,000 acre-feet for development in Arizona. It allocated 600,000 acre-feet for additional irrigation in the Yuma area; 1,100,000 acre-feet to the Central Arizona Project; and the remaining 562,000 acre-feet to future development.

However, under California's diversion-less-return-flows definition, the critical number is the sum of all upstream consumptive uses. The full 2 million acre-feet of Gila River water use upstream, including that made available by salvage by use, fully counts against the apportionments. This has the effect of adding nearly 2 million acre-feet per year to the Colorado River "surplus" at the international boundary.[12] Since California needed surplus water to supply its full 5.362 million acre-feet of contracts, any additional surplus water benefited its agencies.

California's comments on HD 419 primarily focused on the hydrology of the Colorado River and the water supply available for development. California, of course, took exception to Reclamation's use of the term "stream depletion," noting that "the compact does not set 'stream depletion' as a standard for allocations of water but deals with 'beneficial consumptive use' alone."[13]

California then offered its own accounting. Where Arizona had claimed it was currently using 1.4 million acre-feet, California's estimate, based on its diversion-less-return-flows theory, was that Arizona was actually already using 2.3 million acre-feet of water in the Gila River system alone.[14] With the same assumptions about the Upper Basin depletions (7.5 million acre-feet per year) and deliveries to Mexico (1.5 million acre-feet), the effect was to increase the surplus by 1.7 million acre-feet per year. This was enough to largely meet California's 5.362 million acre-feet per year of water contracts but probably not enough water for a Central Arizona Project.

The math is complicated enough in the Lower Basin, what about the Upper Basin? Unfortunately, it is even more convoluted. First we need to back the calendar to the late 1940s and early 1950s. At that time the Upper Basin states had decided that it was to their advantage to be an ally with Arizona and support Arizona's stream depletion theory for measuring apportionments. They did so for several reasons. First, as Hundley claims, it may have bought Arizona's support for the ratification of the treaty with Mexico.[15]

Second, a number of their principle legal and engineering advisors thought that it would allow them to use more water. By measuring their total consumptive use as the net impact of their upstream uses on the flow of the river at Lee Ferry, their compact point, they thought they could consume about 8 million acre-feet per year under a 7.5 million acre-foot apportionment. This was not the two-for-one bargain that Arizona got from the Gila, but it was still a good deal. And, third, California had become their mortal water enemy. California had opposed the Mexican treaty, and California was opposing planning for Upper Basin projects.

The Upper Basin's delivery obligation to Mexico under Article III(c) of the 1922 compact had not yet become a major issue for most of them.[16] Further, assuming the natural flow of the Colorado River at the international boundary was over 17.5 million acre-feet (the best estimate at that time was 17.72 million acre-feet per year) then there was still a sufficient surplus for Mexico (and if California got hurt, that was their problem). What most of the Upper Basin water experts never considered was what would happen if the long-term natural flow of the Colorado River at Lee Ferry continued its downward trend and became less than 17.5 million acre-feet per year. Under the plain reading of the 1922 compact, 17.5 million acre-feet per year is the critical number. If the water available for consumptive use falls below 17.5 million acre-feet, then there is not enough surplus, and the Upper Basin has to provide half of the deficiency. If the water available falls below 16 million acre-feet, then the deficiency is the entire 1.5 million acre-feet required by the treaty, and the Upper Basin's share would be 750,000 acre-feet per year.[17]

Under the situation where the natural flow at the international boundary is less than 17.5 million acre-feet per year, the Upper Basin states are in the identical position that California was in the early 1950s. The additional water made available for Arizona under its stream depletion theory is water that could otherwise be considered surplus. It means that the deficiency under Article III(c) is larger, requiring the States of the Upper Division to make additional deliveries at Lee Ferry on top of their ten-year, 75-million-acre-foot obligation under Article III(d).

Believing that the natural flow of the river at the international boundary was greater than 17.5 million acre-feet per year and that it could increase the water they could develop, the Upper Basin states adopted Arizona's stream depletion theory of accounting, and the historical record suggests it led to an unusual political alliance. The historical evidence is murky as to whether there was an explicit back room political deal cut as early as 1944, in which the Upper Basin states would support Arizona in its conflicts with California, and Arizona would support the Mexican treaty as part of the bargain.[18] But regardless of the political details, the technical issue was clear—both Arizona and the Upper Basin shared an interest in salvage-by-use accounting. In a technical memo to the Colorado Water Conservation Board, Royce Tipton suggests that, indeed, Arizona and the Upper Basin states reached a common technical understanding that they both believed would be in their long-term interests.[19] In the same way that Arizona would benefit from accounting for water use at the confluence of the Gila and the Colorado rather than at the point of diversion and use, the Upper Basin would benefit from accounting for its consumptive use based on its impact on the river's flow at Lee Ferry. The impact of salvage by use in the higher, cooler Upper Basin was less than in the hot deserts of southern Arizona, but it was enough to forge the unusual alliance.

This decision had consequences beyond keeping Arizona as an ally in the Senate battle for the ratification of the Mexican treaty. Arizona's stream depletion theory would be made a part of the 1948 Upper Basin Compact. Arizona became a solid supporter of the Upper Basin's plans for comprehensive development. Finally, as we shall see, it became a factor in the Upper Basin states' decision to stay out of the Supreme Court battle between Arizona and California. Perhaps had the Upper Basin been a little more curious and thoughtful about the long-term average flow of the Colorado River, they might have made different decisions. Better science was available, but like their predecessors in the 1920s, they never asked for it.

Another Missed Opportunity to Rethink the Hydrology

While HD 419 focused the debate on salvage by use with its calculation of the Colorado River's natural flow at the Mexican border, lost in the argument between Arizona's and California's different accounting methods was the underlying reality that there was less water to allocate than the deals of the 1920s had suggested. HD 419 contained a vast list of potential projects. In that sense, it represented the logical completion of the idea E. C. LaRue had pushed from the beginning—not piecemeal dam building, but rather a plan for comprehensive Colorado River Basin water development. But while embracing LaRue's dam-building side, it failed to wrestle with the reality that there was less water and the basin's ambitions needed to be scaled back, continuing the pattern of overly optimistic projections.

HD 419's major contribution to science was its calculation of the natural flow of the Gila, a notoriously difficult task.[20] HD 419 estimated the Gila's average natural discharge into the Colorado River at Yuma at 1.27 million acre-feet per year. Upstream in the Phoenix area, the report's authors estimated the flow of the Gila and its tributaries at 2.29 million acre-feet per year. The biggest failure of the HD 419 hydrology was that it did not advance our understanding of hydrology of the river as a whole.

At a time when the Department of Defense, which considered the Hoover Dam power production used for the Southern California aircraft industry a critical defense asset, was reaching out to the science community to study the long-term flow of the river, the Bureau of Reclamation chose to simply update the mathematical calculation of the estimated mean natural flows at a few points on the river.[21] The estimated annual natural flows for these locations were less than those used by the 1922 compact negotiators and the secretary of the interior for the Hoover Dam water contracts, but they were still much higher than the modern values. The large uncertainties associated with the estimated Lee Ferry flows before 1922 were not adequately disclosed and considered. There was no desire or attempt to improve our understanding of the hydrology for the period from 1878 to 1898. The flow reconstructions made by E. C. LaRue and Herman Stabler and the conservative hydrologic analysis made by the Sibert board were again ignored. Had the HD 419 authors considered the limited information available to them and sought input from the science community to improve it, they could easily have concluded that the period of 1906–30 was unusually wet compared to the overall 1878–1943 period and that the long-term average natural flow of the Colorado River at Lee Ferry was not 16.27 million acre-feet per year as reported by HD 419 but closer to 15 million acre-feet per year.[22]

Every state but California saw in the report a promising potential for more dams and water diversions to support growing agricultural and municipal development. For Arizona, it meant a Central Arizona Project to bring Colorado River water uphill into the state's central valleys. For the states of the Upper Basin, it meant a range of projects that would dam and divert water for agriculture within the basin, and for transbasin diversions to growing population centers outside the basin. Since all but a few of these projects would be built by the Bureau of Reclamation, its senior managers had the same basic motive as did Arthur P. Davis in 1922 and E. B. Debler in 1930: optimistic hydrology meant more work for their agency.

Like 1922 and 1929, the states were not all in agreement. Now it was California, which viewed HD 419 and the projects it promoted as a threat to the projects it had already completed, including the 1.2 million acre-feet a year needed to fill its Colorado River Aqueduct and supply its growing coastal cities. While the players argued in ways intended to claim and protect their share of the river's water against perceived enemies across state and basin boundary lines, they once again failed to explore and recognize the heart of their problem—the gap between their aspirations for development and the Colorado River's hydrologic reality.

CHAPTER 13

The 1948 Upper Colorado River Basin Compact

In the summer of 1948, as the Upper Basin states, along with Arizona, gathered to divide the Upper Colorado River Basin's waters, one of the most important flaws in the 1922 Colorado River Compact had become clear. The allocation of a fixed supply of water, 7.5 million acre-feet per year for the Upper Basin and 8.5 to the Lower Basin, on a river with less water than the compact's framers had thought, had become a problem.

Article III(a) of the 1922 compact apportions 7.5 million acre-feet per year of beneficial consumptive use to the Upper Basin. Arizona's portion of the watershed was small but was enough to give it an important seat at the table as the 1948 deal was being struck. The intent of the 1922 compact negotiators was that the two subbasins would complete the job of deciding how much went to each state. This created an opportunity to fix some of the problems created a quarter century earlier by the compact's overallocation of water. The response of the states' representatives—to allocate water by percentage rather than absolute amounts—reflected a major advance in integrating Colorado River water policy with hydrologic reality, providing a built-in mechanism to respond to variability. During dry times under a percentage rather than absolute allocation scheme, everyone would share cutbacks equally. But the authors of the Upper Basin Compact could not make a clean break from the problems of the past. To make the difficult deal they were negotiating palatable, they continued the long-standing tradition of assuming an unrealistically large supply of water.

Preparations for an Upper Basin Compact began in the 1930s. After World War II ended and the Mexican treaty was ratified, the effort became focused. The compact had two major legal purposes and a primary political purpose. The legal purposes were, first, to divide the water apportioned to the Upper Basin by the 1922 compact among the five states and, second, to determine each state's obligation in the event that water use had to be curtailed in order to meet their obligations to the Lower Basin under the 1922 compact. The political purpose was to align the Upper Basin states behind a unified comprehensive strategy for development they could implement through federal legislation.

The Compact Negotiations

Unlike in 1922, the 1948 compact parties spent considerable time preparing for the negotiations. In 1929, Delph Carpenter prepared a conceptual draft of an Upper Basin compact. His proposal was relatively simple and true to his principle of state sovereignty. He suggested that each state with Upper Basin interests could use the Colorado River water that arose within that state without interference from the other states, only subject to the obligations of the Upper Basin under the 1922 compact.[1] This concept greatly favored Colorado, which produced over 70 percent of the Upper Basin yield, but disadvantaged New Mexico, which produced only about 2 percent. There is little evidence the other states took the proposal seriously. However, it did set up one of the critical questions that would face the negotiators: How much water would Colorado have to concede to the other states in order to reach a unanimous agreement?

In 1934, the Upper Basin states met in Denver and agreed that development could temporarily proceed without a compact subject to the concept that each of the four states could veto development in the other states if a state believed that the proposed development might exceed the amount of water the sponsoring state would obtain under a compact. This concept allowed development of projects like the Colorado–Big Thompson to proceed.[2] In contrast to what was happening in the Lower Basin, the 1934 agreement set the stage for Upper Basin collaboration that would carry the states through a successful compact and keep them aligned on major federal legislation and interstate litigation. This alliance remains strong today.

The states and the Bureau of Reclamation also agreed to spend considerable resources for studying and determining the hydrologic facts. In 1938, Royce Tipton completed a report on the available water supply.[3] Tipton concluded that the natural flow of the Colorado River at Lee Ferry was 16.4 million acre-feet per year. He referenced studies by LaRue, Debler, and Jacobs and Stevens and concluded, "With respect to the water supply of the Colorado River, authorities using necessarily, the same basic data, but applying different methods, have arrived independently at conclusions which are in close agreement."[4] Although he used LaRue as a reference and made note of the river stage readings at Yuma beginning in 1878, Tipton made no mention of either Stabler's or LaRue's pre-1900 flow reconstructions or the conclusions of the Sibert board, thus again ignoring the 1870s–1905 dry period.

As the compact negotiations proceeded, the Upper Basin's engineering committee concluded that the long-term average natural flow at Lee Ferry was 15.64 million acre-feet per year, based on a period of record of 1914–45. Yet again, the average and thus the amount of water available for use continued its downward trend. And yet again, it ignored the 1870s–1905 dry period and the warnings of LaRue, Stabler, and the Sibert board.

As in 1922, the 1948 compact commissioners elected the federal commissioner, Harry Bashore, as the chairman. Bashore, a retired commissioner of the Bureau of Reclamation, brought considerable expertise and skill to the commission. As the federal commissioner, Bashore's goal was to advance Reclamation's construction agenda to build the projects identified in HD 419. As with the 1922 negotiations, each commissioner brought with him a number of advisors. The Colorado team included consulting engineer Royce Tipton. Tipton had been an important participant in the negotiations and ratification of the treaty with Mexico and would greatly influence the direction of the negotiations.[5]

From July 1946 through October 11, 1948, the commission held eight formal meetings. Most lasted several days, several more than a week. In late 1948 and 1949, it held three more meetings to address ratification issues.

Unlike the 1922 proceedings, the Upper Basin commissioners had no legislatively mandated deadline. However, they all understood that an Upper Basin Compact was a prerequisite to comprehensive water development in the basin. This added a sense of urgency. The first four meetings focused on a discussion of the factual background and legal structure of the compact, refining issues, and making assignments to the legal and engineering committees. During meetings five, six, and seven, they made the major decisions that would shape the compact. Meeting eight was dedicated to final drafting.[6]

Questions concerning the basin hydrology and how to address them in an Upper Basin compact were at the top of commission's list. A natural flow at Lee Ferry of 15.64 million acre-feet per was equivalent to a 17-million-acre-feet-per-year natural flow at the international boundary with Mexico. This was a full 1 million acre-feet per year less than the 18 million acre-feet per year that was assumed available during the Senate hearings on the treaty with Mexico just a few years before. More importantly, with only 17 million acre-feet per year available for use from the Colorado River, not only was there no surplus, the water apportioned under the 1922 compact and the 1944 treaty totaled 17.5 million acre-feet; therefore, the river was now, on paper, overapportioned.[7] For the Upper Basin, it meant that during decades like the thirties, there would be less than 7.5 million acre-feet of water available. With no certainty as to how much water was available, an apportionment of a fixed amount of water to the individual states would be problematic. During the fifth meeting the commission, Colorado legal advisor Jean Breitenstein pointed out the problems created by the 1922 compact negotiators' decision to allocated fixed amounts of water. "Now we come along twenty odd years after the Colorado River Compact and hear the engineers say perhaps there wasn't as much water in the Colorado River as the negotiators of the Colorado River Compact thought there was. That immediately raises the point as to whether or not we shall fall into the same error, if it was an error."[8]

However, using percentages created complications. The commissioners now needed to clearly define the "what" the percentages would be applied to. Without a clear understanding of the amount of the "what," it would be hard for states to plan their development and for anyone to know if a state (or states) were using more than its (or their) apportionment. The commission ultimately defined the "what" as "the total quantity of consumptive use per annum apportioned in perpetuity to and available for use each year by the Upper Basin under the Colorado River Compact."

At this point, however, the seemingly realistic discussion of the basin's hydrology took a detour. Rather than asking their engineering committee to determine the amount of water that would likely be available under their definition of the "what," politics appears to have prevailed, and the negotiators turned to the marquee issue of how much each state would get. For this discussion, they reverted to the practice of overstating the river's yield. If the percentages apportioned to each state were applied to a bigger water availability number, then the compact would be much easier to sell back home.

The Engineering Committee had determined the historical and natural water contributions of each state at the state lines and at Lee Ferry. The data show the haves and have-nots. Colorado, with 35 percent of the Upper Basin drainage area, contributed over 70 percent of the natural flow at Lee Ferry. New Mexico, with 8 percent of the drainage area, contributed just 1.58 percent of the Lee Ferry flow. The aggregate numbers were the reason for the Upper Basin's focus on salvage by use and Arizona's stream depletion theory of accounting.

The sum of the natural flows at the state lines was 16.2 million acre-feet per year, but the natural flow at Lee Ferry was 15.6 million acre-feet per year, 600,000 acre-feet less. These 600,000 acre-feet were the basis of the technical bond that held Arizona and the Upper Basin states together. Royce Tipton believed that like the Gila River, the Upper Basin could also take advantage of the natural losses (or "wasting," as they called it then). If the Upper Basin Compact defined a depletion for apportionment purposes as the impact of human water use to the natural flow at Lee Ferry (the Upper Basin version of Arizona's stream depletion theory, which had been so important in its attempt to win the accounting battle over the use of Gila water), then by consuming the water upstream, the Upper Basin could, at least during wet cycles, collectively consume 600,000 acre-feet more than 7.5 million acre-feet per year and, under its 1922 compact apportionment, a total of 8.1 million acre-feet.[9]

At the beginning of the seventh meeting Tipton made a detailed presentation on his concept of salvage by use and why it was to the Upper Basin's benefit to measure the consumptive use of an individual state as the impact of a state's human depletions on the natural flow at Lee Ferry. He was assisted and supported by Arizona engineering advisor R. I. Meeker.[10] Meeker told the commissioners and their numerous advisors that he had been present during all of the Santa Fe meetings in 1922. It was his opinion that the 1922 compact negotiators intended that the 7.5 million acre-feet of apportionment provided to the Upper Basin be measured as impacts to the natural flow at Lee Ferry. Tipton went on to make a recommendation as to how the commission could actually calculate consumptive uses at Lee Ferry by what he referred to as the "inflow-outflow" method. Initially, the idea of measuring consumptive use at Lee Ferry was opposed by Wyoming.

Wyoming legal advisor William J. Wehrli opposed Tipton's recommendation. He believed that by the use of the term *beneficial consumptive use* in the 1922 compact, it was not clear the apportionments were to be measured at Lee Ferry for the Upper Basin and at the international boundary for the Lower Basin.[11] He also noted that under Tipton's interpretation the Lower Basin would benefit by about 1.5 million acre-feet compared to an Upper Basin benefit of 400,000–600,000 acre-feet. Further, it allowed the Lower Basin to consume 9 million acre-feet under its 7.5-million-acre-foot apportionment (or 10 million acre-feet under its 8.5-million-acre-feet III [a] and [b] total). Importantly, it would decrease the available surplus, thus increasing the Upper Basin's burden to Mexico under the 1922 compact. Finally, the Wyoming delegation understood and was uncomfortable with the idea that the Upper Basin states were taking sides with Arizona in its dispute with California.[12] The commission was initially unable to resolve the differences between Wyoming and Colorado, so it went on to the signature issue of how much water, or what percentage, each state would receive under an Upper Basin compact.

The apportionment negotiations dominated the marathon seventh meeting.[13] Colorado opened the bidding by suggesting it would accept an apportionment of 56 percent. However, it qualified the offer based on the other states accepting four conditions—Tipton's two recommendations concerning measuring consumptive uses at Lee Ferry and using the inflow-outflow method and two more recommendations by the Legal Advisory Committee. The first concerned the accounting of reservoir evaporation, and the second involved the obligations of each state during a curtailment. The other states presented their initial offers and their rationale. The total requests for the four Upper Basin states were 115 percent.[14] In justifying their individual percentage requests, some of the states took advantage of additional water made available through salvage by use.[15]

To move the negotiations forward, Chairman Bashore and Colorado's Stone focused on the potential development of each of the states as presented by Reclamation's engineering reports in HD 419. The Reclamation report showed potential development in the Upper Basin of 9.136 million acre-feet.[16] This was more than what was available, but it could be used as a basis to determine each state's percentage apportionment. This approach narrowed the problem but didn't resolve the states' differences. On July 19, the commission formally convened for an eight-minute meeting where Bashore made a recommendation for a settlement. The states then caucused and, a day later, accepted the Bashore numbers.[17] For the formal vote, they used a paper ballot. Bashore reasoned that with a paper ballot none of the individual commissioners could be accused by their home states of being the first to blink. Table 6 shows the states' initial requests and the final apportionments that became the basis for Article III of the compact.

As one can see, Arizona accepted a small fixed amount of water, which, given it had no flow obligations at Lee Ferry, made administration of the compact much simpler. The big winner among the four Upper Basin states was New Mexico. Commissioner Fred Wilson successfully argued it was needed because his state included potentially large water rights claims by the Navajo Nation and the Jicarilla Apache Tribe.[18] The commission considered two basic approaches to addressing Indian water rights. The first was to take the Indian water off the top and apply the state percentages to the remainder. The second was to include Indian uses in each state's apportionment. They chose the latter.

Under the Winters Doctrine, established by the U.S. Supreme Court in 1908,[19] the Native American rights would likely be senior to the other projects that served non-Indian uses. Because the Upper Basin Compact negotiators intended the apportionments to cover all existing and future uses, if the New Mexico legislature was going to ratify the compact, it needed an apportionment large enough to cover both Indian and non-Indian uses.[20]

Table 6 Upper Basin Compact initial requests and final apportionments

State	Percent of initial request	Percent of final apportionment	Percent contributed at Lee Ferry
Colorado	56	51.75	70.14
New Mexico	15	11.25	1.58
Utah	28	23	16.38
Wyoming	16	14	11.03
Subtotal	115	100	99.13
Arizona	140,000 acre-feet	50,000 acre-feet	00.87

Perhaps because Colorado agreed to further reduce its request to 51.75 percent, Wyoming dropped its opposition to Tipton's definition of consumptive use and his inflow-outflow methodology allowing an agreement among all of the parties.[21] Article VI of the 1948 compact provides that "the Commission shall determine the quantity of the consumptive use of water, which is apportioned by article III hereof, for the Upper Basin and for each Upper Basin state by the inflow-outflow method in terms of man-made depletions of the virgin flow at Lee Ferry, unless the Commission, by unanimous action shall adopt a different method of determination."[22]

There was also consensus on how to determine each state's Lee Ferry obligation in the event of a curtailment. The problem was how to draft the provision. Further, the commission decided that overuse was not a problem except when an overuse by one state would impact another state, such as when that overuse was the cause of an actual curtailment.[23]

Article IV, which addresses the subject of a curtailment, provides that if a state or states exceeded its consumptive use apportionment percentage as measured by the total use over the preceding ten years, then that state (or states) has to deliver at Lee Ferry the aggregate of its total overdraft before any demand is made on any other state. If all four states were using their actual apportionments for the previous ten years, or once the overdrafting states have paid back their ten-year penalty, then each state has to provide an amount of water based on its post-1922 compact uses in the year prior to the curtailment.[24]

To implement these last two provisions, the compact created the Upper Colorado River Commission, charged with making decisions required by the compact, such as reservoir evaporation accounting and formally determining the amount of any deficit at Lee Ferry needed for compliance with Article III of the 1922 compact and dividing water deficits up among the four Upper Division states.[25]

Arizona's allocation in the final agreement was small, but it came away from the deal with something much more important. The state's representative, Charles A. Carson Jr., had come to the negotiations with two goals. Obviously, he wanted to obtain a bit of water for Arizona's Upper Basin lands. More importantly, he wanted to maintain Arizona's close working alliance with the upper states. In the few short years since Arizona had ratified the compact, it had transformed its relationship. It had gone from Delph Carpenter's nightmare rogue state bent on undermining the 1922 compact to a close ally of Upper Basin leaders Clifford Stone and Royce Tipton, helping obtain Senate ratification of the 1944 treaty with Mexico. Carson knew that maintaining this alliance was essential to Arizona's future plans, and the final deal did that. By coming to agreement on a method for accounting water depletions that was mutually beneficial, he had created an alignment between Arizona and the Upper Basin states that would have lasting impact.

Summary and Analysis of the Upper Colorado River Basin Compact

The final drafting of the Upper Colorado River Basin Compact was completed during the eighth meeting of the commission and signed on October 11, 1948, in Santa Fe, New Mexico. In many ways, the 1948 compact is considered an improvement over the 1922 compact in both process and substance. The commissioners were more prepared and more deliberate. They had available to them a detailed engineering report. The commission's formal meetings were open to the press. And, although California opposed the provisions of the compact that supported Arizona's stream depletion theory, the five signatory states and ultimately Congress quickly ratified the compact within a year of it being signed.

The 1948 compact is much longer and more complex than the 1922 compact. It includes twenty-one articles. The critical ones are Article I, which defines the purpose and makes it subject to all provisions of the 1922 compact; Article III, which includes the apportionments; Article IV, which describes the procedures for a curtailment; Article V, which addresses reservoir evaporation and reservoir accounting; Article VI, which defines consumptive use as measured at Lee Ferry; and article VIII, which creates and defines the powers of the Upper Colorado River Commission.

As expected, during the congressional ratification process, California opposed how measuring consumptive use was defined by Article VI.[26] In its ongoing dispute with Arizona, California took the position that its diversions minus return flows theory was the proper method of measuring apportionments under the 1922 compact and 1928 act. It believed that Arizona would use the congressional approval of the 1948 compact to make the legal case that Congress was endorsing Arizona's stream depletion theory of apportionment accounting. This opposition resulted in a spirited discussion and debate among the witnesses from the Upper Basin and representatives from California during the ratification hearings held by the House Subcommittee on Irrigation and Reclamation. Ultimately the five signatory state representatives and the subcommittee agreed to report language that "congressional consent does not, nor does the Upper Colorado River Basin Compact, alter, amend, modify, or repeal the Boulder Canyon Project Act, or the Colorado River compact."[27] The purpose of this language was to maintain federal government neutrality in what everyone expected next, a Supreme Court battle between Arizona and California.

CHAPTER 14

Upper Basin Development Moves Forward

At a House hearing in the spring of 1955, California Republican Craig Hosmer unloaded:

> The expressed philosophy of the Bureau of Reclamation embraces the contention that reclamation is all things to all men. As it has done year after year in the past, the Bureau has come forward again with a thesis attempting to illustrate the great benefits to be derived to the nation from a western irrigation project. The subject this time is the multi-billion-dollar Colorado River Project. It is the Bureau argument that if the doors of the Federal Treasury are opened to the proponents of this fiscal monstrosity, every state in the union will get some of the loot.... When it comes to economists, the Bureau is woefully deficient. The economics of the Bureau of Reclamation are as un-sound and as phony as a three dollar bill.... The Bureau's policies have been called creeping socialism. I submit that the Bureau's economics wouldn't be tolerated by the most ardent Socialist. And they are certainly not creeping policies. They are advancing with the speed of a jet plane.[1]

The occasion was an appearance by Reclamation commissioner Paul Dexheimer in what was mostly friendly territory. The subcommittee, under the chairmanship of western Colorado's Wayne Aspinall, was stacked with westerners anxious for irrigation and municipal water delivery projects his agency would build. This is how the "iron triangle" was supposed to work—Reclamation offering up water development projects members of Congress needed to satisfy constituents back home. On that day, Dexheimer and his lieutenants would be testifying about projects in the Upper Colorado River Basin, Aspinall's home territory. But by the mid-1950s, whatever détente had existed among the states in support of a common vision of Colorado River development was near complete breakdown. It was becoming increasingly clear that there was not enough water to meet the desires of all the iron triangle's players.

There had already been signs—in California's opposition to a Mexican treaty, and the struggles over the plans laid out in the ambitious House Document 419. But what had been subtle obstruction in those cases by the mid-1950s has become open political warfare. California already had its big water projects, including enough infrastructure to use nearly 1 million acre-feet of surplus water beyond its nominal 4.4-million-acre-foot allocation. But the inexorable math of the Colorado River was catching up with development efforts. Realizing that with additional Upper Basin water use the surplus on which it had come to depend would not reliably be there, California fought hard.

Less than three decades after the Boulder Canyon Project Act's passage, the failure to heed LaRue, Stabler, and Sibert had become an obstacle to the river's further development. Yet as Hosmer thundered about "creeping socialism," the real problem—too little water—remained largely unsaid.

The 1956 Colorado River Storage Project Act

Hosmer's testimony came as Congress was considering the next critical steps in Colorado River development. After the five signatory states and Congress ratified the Upper Colorado River Basin Compact, the Upper Basin states pursued federal legislation to provide for the comprehensive water development of the Upper Basin. They had two primary objectives. First, they needed reservoirs capable of storing 30 million acre-feet of water. This would provide enough storage, filled during wet times, to weather a repeat of the

1930s drought and still meet the Upper Basin's obligations under Article III of the 1922 compact. Second, they needed a way to subsidize the development of the irrigation projects that would use the major portion of the Upper Basin's compact share, because it was clear farmers could not afford the cost. "The only source of revenue on the Colorado River storage project," Reclamation engineer C. B. Jacobson told members of the Upper Colorado River Commission in 1952, "is from power."[2]

Glen Canyon Dam was the lynchpin. Its 25 million acre-feet of capacity would provide most of the hold-over storage, and its hydroelectric power would provide the subsidies. After the Bureau of Reclamation completed a planning report on the Colorado River Storage Project in 1951, it was up to the states to convince Congress. This was not simple. California supported the need for hold-over storage, but strongly opposed using power revenues to subsidize other Upper Basin projects. Its Republican allies considered the proposal, as Hosmer had described it, "creeping socialism." After nearly five years of congressional debate, in 1956 the Colorado River Storage Project Act (CRSPA) passed and was signed into law by President Dwight Eisenhower.[3] The 1956 act authorized the construction of Glen Canyon Dam, which creates Lake Powell and three more large upstream storage reservoirs and put in place a method for subsidizing what were referred to as "participating" projects.

By the 1950s, the approach in the Upper Basin had evolved to one where the 1922 compact was far more than Carpenter's view of a water treaty among states defining rights, obligations, and opportunities. The water available to the Upper Basin states under the compact was now considered an entitlement. If these states had neither the economic need nor the financial resources to put their compact water to beneficial use, then the federal taxpayers owed the basin the resources to do so.

The idea that the federal government would subsidize the irrigation components of Reclamation projects was not new. The repayment contracts for project irrigation features built by the Bureau of Reclamation normally had fifty-year terms at zero percent interest rate—preserving the fiction that irrigators were reimbursing the government while effectively providing an enormous subsidy.[4] Further, costs attributable to flood control, recreation, and fish and wildlife were considered national benefits and nonreimbursable. CRSPA went beyond this in several ways. First, the act specifically provided that revenues from the sale of hydroelectric power generated by the storage projects would be used to repay the federal treasury for capital costs of irrigation projects that the secretary determined were beyond the ability of an irrigator to pay. Second, the repayment, with no interest, could occur at any time during the fifty-year repayment period. This allowed Reclamation to schedule the payments in the final few years of the fifty-year period, which both minimized the impact on the power rates and maximized the federal subsidy. Under the act, many of the individual irrigation projects (participating projects) that would receive the power subsidy had to be further authorized by Congress.[5]

While California's overt opposition to the CRSPA centered on the financial issues, the underlying objection was water supply. By the early 1950s, when CRSPA was first proposed, California was using over 5 million acre-feet of Colorado River water under its 1930s contracts for Hoover Dam water. It had lost the battle in Congress over the ratification of the 1944 treaty. It had used its political power to block the congressional authorization of the Central Arizona Project. Arizona and the Upper Basin states had successfully completed the Upper Colorado River Basin Compact, where the compact parties agreed to Arizona's stream depletion theory of accounting. In March 1954, when the Eisenhower administration decided to support CRSPA, giving the legislation an important boost, Arizona and California were already locked in battle in the U.S. Supreme Court.

California's math was straightforward. With a little more than 15 million acre-feet per year of natural flow at Lee Ferry and, with the help of CRSPA, the Upper Basin consuming 6 million acre-feet, there were 9 million acre-feet of water left for uses on the Lower Basin main stem. With Mexico getting 1.5 million, and the tributary inflow between Lee Ferry and Hoover Dam offsetting reservoir evaporation, it meant there was about 7.5 million acre-feet of main stem water available for three states. This was enough for California's 4.4 million acre-feet, but not enough for a surplus. Thus, it had no choice but to oppose federal legislation that would give the Upper Basin states the resources to accomplish their water development goals. Politically, it meant California against the rest of the basin.[6]

Even with strong economic arguments against the project and the Bureau of Reclamation's horrible record of building projects for anywhere close to their projected cost, the opponents never had much of a chance. President Eisenhower supported the bill. Texas Democrat Sam Rayburn, the powerful speaker of the U.S. House of Representatives, supported the bill. There were even supporters from California. The House Interior Committee was then chaired by Representative Claire Engle, who supported the bill because he had a project in his Northern California district, the Trinity River project, that he wanted approved.

By reaching as far as the Trinity River, the "iron triangle" held. The Senate passed CRSPA in late 1955. The House of Representatives debated the bill for three days in late February 1956. Colorado's Aspinall led the supporters. California's Craig Hosmer led the opposition. Supporters focused on the value of the project to the region and the nation. Underlying their argument was the question of equity. The federal government had built great projects for the Lower Basin. With the CRSPA, supporters argued, it was doing the same for the Upper Basin. The myth that the 1922 compact had divided the waters about equally between the two basins fed that argument.

On March 1, 1956, the House version of CRSPA passed the House on a vote of 256–136. A conference committee quickly ironed out the differences between the House and Senate versions. The Colorado River Storage Project Act was signed by President Eisenhower and became effective on April 11, 1956.

From the Upper Basin's perspective, CRSPA was well timed. Its passage came during the second major drought of the twentieth century. The 1953–64 drought was drier and longer than the 1930s drought. More importantly, for the thirty-four-year period of 1931–64, the average natural flow at Lee Ferry was slightly less than 13.5 million acre-feet per year and the average natural inflow to Lake Mead was less than 14.3 million acre-feet per year. The Sibert board hydrology that was considered too conservative and easily dismissed in 1928 now looked like the rule, not the exception.

Development Moves Forward, Then Stalls

Once CRSPA became effective, the Bureau of Reclamation moved quickly. The act authorized four initial units: Glen Canyon Dam (Lake Powell), Flaming Gorge Dam, the Curecanti Unit (now named Aspinall, and which consists of three dams[7]), and Navajo Dam. Since these big reservoirs, especially Glen Canyon, were the cash registers for the participating projects, Reclamation began construction as soon as the final designs and project specifications were completed.[8] Construction on Glen Canyon Dam began in October and was topped out in September 1963. Its turbines began generating power in September 1964. Unlike Hoover, filling Lake Powell was extremely complicated, both politically and hydrologically.[9] The challenge was complicated by the fact that other initial units, including Flaming Gorge with its capacity of nearly 4 million acre-feet, would be filling at the same time. Filling Lake Powell for the first time would take seventeen years.

CRSPA authorized eleven participating projects for immediate construction. After further feasibility and economic studies, two more projects were added in 1962, three in 1964, and five in 1968. The strategy of Upper Basin states to use the participating units to develop a large portion of their compact share was only partially successful. Four of the authorized projects were found economically infeasible and never built. Many more were downsized, and on two projects, the irrigation components were completely eliminated.[10] The problems they encountered in developing successful irrigation projects in the Upper Basin would not have surprised Delph Carpenter. He fully understood the limitations on Upper Basin agriculture, and as a proponent of state sovereignty, he would have been shocked by their reliance on federal resources.

Agriculture was not irrelevant to the future of the Upper Basin. In Colorado, for example, total irrigated acreage doubled from the mid-1950s to the early 1980s.[11] The addition of subsidized agricultural water via the Colorado River Storage Project and other similar efforts had an important role to play. "Development" was not some abstract concept—people were really putting the water to use. But with the higher elevations, mountainous terrain, and shorter growing season, the opportunities were far less expansive that those available to the Lower Basin states.

While irrigated agriculture in the Upper Basin never reached the scope the early boosters had hoped, the states of Colorado, Utah, and New Mexico found other ways to use much of their water by diverting it to municipal areas and farms outside the Colorado River Basin. The Central Utah Project moves water from the Duchene River to the Wasatch Front and the San

Juan-Chama Project diverts water into the Rio Grande basin for use in New Mexico. These federal projects were supplemented by projects funded by the region's larger cities, such as Denver, Aurora, and Colorado Springs.

Because of CRSPA (both evaporation on the storage reservoirs and the consumptive use from participating projects), nonfederal municipal development (mainly exports), and the construction of thermal electric power plants, consumptive uses in the Upper Basin doubled from about 2 million acre-feet per year in the late 1940s to about 4 million acre-feet per year by the late 1980s.

At which point development plateaued.

CHAPTER 15

Arizona v. California

There is an adage among lawyers that "bad facts make bad law." By the 1950s, the "bad facts" that came from ignoring LaRue, Stabler, and Sibert—the disconnect between the paper allocations of the Law of the River and the wet water of the actual Colorado—were driving the basin toward bad law. The forum was the U.S. Supreme Court. The issue at hand was not science but how to come to terms with the reality that the laws written in the previous three decades failed to say what should happen if there was not enough water.

As Upper Basin development moved forward, the conflict between Arizona and California was a stalemate. California's powerful congressional delegation was blocking the authorization of the Central Arizona Project (CAP), arguing that under the 1922 compact the project lacked a sufficient legal water supply. This left Arizona no choice but to take its fight to the U.S. Supreme Court. Arizona filed its suit in 1952. It took the Supreme Court until 1963 to render a decision. Perhaps in an effort to avoid inflicting a fatal blow to the 1922 compact, the court limited its decision to interpreting the intent of Congress under the 1928 Boulder Canyon Project Act. Once the court took this direction, California's fate was sealed, and Arizona stumbled to a victory. The winners were Arizona and its CAP, the Indian Tribes with main stem reserved water rights, and the power of the secretary of the

Interior. The losers were the California agencies with Hoover Dam contracts and, perhaps, because they failed to understand the basic mathematics of the 1922 compact or perhaps because they believed their participation in the case would threaten their development plans, the Upper Basin.

The Basic Mathematics of the Dispute and Its Origins

Arizona's decision to file its 1952 case was the product of a series of falling dominoes, bad decisions not supportable by the reality of the river's hydrology. Each domino was directly or indirectly the result of basic failures of the negotiators and promoters of the 1922 compact, the 1928 Boulder Canyon Project Act, and the Hoover Dam water contracts to fully understand the amount of water available in the river.

The first domino was Article III(b) of the 1922 compact, which allowed the Lower Basin to increase its beneficial consumptive use by 1 million acre-feet per year, giving it a total apportionment of 8.5 million acre-feet per year. There is no question Article III(b) exists because of the tenacity of Arizona's Winfield Norviel, but just about everything else about the article is a matter of dispute.[1] It is commonly believed its purpose was to allow Arizona to fully develop the use of the waters of the Gila River, but the plain language of the compact apportions it to the Lower Basin in general, not specifically to Arizona. Hoover had testified that III(b) water could be developed on either the Lower Basin tributaries or the main stem.[2] This opened the opportunity for an argument over how apportionments under Article III should be measured. Did the compact negotiators intend that the 7.5 million acre-feet apportionments under III(a) and the 1 million acre-feet apportionment under III(b) be measured as the sum of the consumptive uses at the locations of the use (California's diversions-minus-return-flows theory) or the net impact of the depletions to the Colorado River main stem at the international boundary (Arizona's stream depletion theory)? In simple terms, should Arizona's 2 million acre-feet per year of total consumptive uses on the Gila River count as 1 million acre-feet of apportionment or 2 million acre-feet of apportionment?

The next domino was the 1928 Boulder Canyon Project Act. As a prerequisite to a six-state compact that did not include Arizona, the Upper Basin states insisted on language limiting California's use. The act did so by limiting California to 4.4 million acre-feet per year of Article III(a) water plus half of the unapportioned surplus. It also suggested a Lower Basin compact where, in addition to California's allocation, Arizona would receive 2.8 million acre-feet and Nevada 300,000 acre-feet of III(a) water. But in doing so, the 1928 act ignored the 1 million acre-feet of Article III(b) water and failed to define *surplus*. This added to the conflict between Arizona and California.

The next domino was the Hoover Dam water contracts. Based on an assumption that a long-term firm supply of 10.5 million acre-feet per year of water was available for delivery below Lake Mead, the secretary entered into 5.36 million acre-feet per year of contracts with California agencies. This was followed by the 1930s drought, which cut in half the surplus the 1922 compact negotiators thought available for Mexico and future apportionments.

The 1940s dominoes were Arizona's ratification of the compact, its signing of a water supply contract for 2.8 million acre-feet per year of water from Lake Mead, the treaty with Mexico that squeezed the available surplus to as little as 200,000 acre-feet per year, and the shared understanding among Arizona and the Upper Basin states on the use of the stream depletion theory of accounting.

The final dominoes were the completion of a feasible plan for the development of the CAP by the Bureau of Reclamation and California's ability to block the authorization of that project in Congress.

Thus, as the 1950s began, Arizona and California were in similar positions. Each had plans for or current uses that exceeded 5 million acre-feet per year of Colorado River water, but legal uncertainty suggesting a risk that their actual supply under the law might be far less. Further, with each year that passed, it was becoming more obvious that the river's long-term reliable supply was far less than the 17.5–18.1 million acre-feet per year at Lee Ferry the compact negotiators thought available.

Arizona's approach was to try to convince the Supreme Court that under the 1922 compact and 1928 act it had a legal apportionment of 3.8 million acre-feet per year, and under its stream depletion theory of accounting, its

2 million acre-feet-plus of consumptive use on the Gila River was only 1 million acre-feet of apportionment. It could just about make the math work.

California's defenses seemed equally plausible. It rejected Arizona's stream depletion argument, claiming Arizona's full 2 million acre-feet of consumptive use on the Gila counted against its 2.8 million acre-feet of Article III(a) water. Further, since Congress only apportioned the 7.5 million acre-feet per year of III(a) water in the 1928 act and ignored the extra million acre-feet of III(b) water, the III(b) water was a part of the surplus, and through the Hoover Dam contracts its agencies had appropriated most of the surplus. Under California's case, the water available for the CAP would be so limited as to make the project infeasible, protecting the water supply to California's contractors.

The Legal Battle

The battle over the legal water supply available first began in Congress shortly after the transmittal of House Document 419. Led by Senator Carl Hayden, Arizona sought congressional authorization of the Central Arizona Project. In 1949, a bill authorizing the project passed the Senate. It was blocked in the House. The lack of legal certainty over the amount of the available water supply for a very expensive project was too much of an obstacle. After Congress debated different approaches to resolve the disputes without success, Arizona filed the lawsuit in August 1952. The court accepted the case in January 1953. The case was ultimately decided in June 1963 and a decree issued in April 1964. Arizona won, but its victory came with significant consequences to the entire basin.

Arizona's 1952 bill of complaint included eight claims for relief. Five of those directly addressed unresolved 1922 compact questions related to the hydrology of the Colorado River:

- Arizona sought a court decree giving it title to 3.8 million acre-feet of
  Colorado River water under Articles III(a) and (b) of the 1922 compact,
  2.8 million acre-feet of main stem water under Article III(a), and all of
  the 1 million acre-feet of water apportioned to the Lower Basin under
  III(b), subject only to the rights of New Mexico on the Gila River and
  Utah on the Virgin River.

- Arizona sought to limit California to no more than 4.4 million acre-feet per year. Arizona believed the III(b) water was not a part of the surplus and that California had no right to any III(b) water.
- As to surplus waters unapportioned by the 1922 compact, Arizona asked the court to decree that California be entitled to one-half and Arizona the remainder less one twenty-fifth for Nevada and whatever rights New Mexico and Utah might have in the surplus.
- Arizona asked the court to find that its stream depletion theory was the proper method of measuring apportionments under the 1922 compact.
- Arizona asked the court to find that losses in and from reservoirs in the Lower Basin on the main stem be charged against the States of the Lower Division in the same proportion as their consumptive use in the Lower Basin. This would charge the lion's share of evaporation to California.[3]

The combined effect of these claims would give Arizona the right to consume about 5 million acre-feet per year of Colorado River water. Arizona could fully consume all of the Gila River (except New Mexico's small share), build the proposed Central Arizona project to move 1.2 million acre-feet of main stem water into Central Arizona, and cover its share of evaporation all within a 3.8 million acre-foot apportionment.[4] Arizona's claims would limit California to 4.4 million acre-feet per year. However, because California's share of evaporation losses would be about 600,000 acre-feet per year, its net usable water would only be about 3.8 million acre-feet. That was about 1.5 million acre-feet per year less than what it could use under its 1930 contracts and the capacity of the projects it had already built.[5]

California's response offered three affirmative, but confusing, defenses.[6] First, it argued that California and its agencies had legal rights to the beneficial consumptive use of 5,362,000 acre-feet per year of waters of the Colorado River system under the Colorado River Compact, the Boulder Canyon Project Act, a "statutory compact" between the United States and California, and the Hoover Dam water contracts. California used the term *statutory compact* to define the legal relationship it had with the secretary of the interior under the 1928 act, which authorized the six-state ratification of the 1922 compact provided that California limited itself to 4.4 million acre-feet per year plus one-half of the unapportioned surplus. It argued that under the 1928 act

Congress had given the basin two choices: a seven-state compact with Arizona as a party and under which any limitations on California would be the subject of a negotiated Lower Basin compact, or a six-state compact, under which Arizona would choose not to be a party, and California would have to agree to limit itself to 4.4 million acre-feet per year plus one-half of the surplus. This arrangement, California argued, amounted to a statutory compact.

This set up California's second defense. Since Arizona chose not to ratify the 1922 compact until 1944, it was prohibited from challenging again in 1952 the 1928 actions it had already challenged without success in the 1930s round of Supreme Court litigation, California's lawyers argued.

California's third line of defense fell back on the doctrine of prior appropriation. California's water rights associated with its 5,362,000 acre-feet of contracts with the secretary of the interior dated to June 1929. The water under those contracts was senior to any Arizona water from projects that had come after or might be built in the future. California argued that its diversions-minus-return-flow theory of accounting was the appropriate way to measure apportionments and that the 1928 act requires it. Again citing the 1928 act and its contracts, California claimed reservoir evaporation and system losses were not to be charged against state apportionments.[7]

Under California's arguments, using diversions minus return flows to measure apportionments, Arizona was using about 2 million acre-feet per year on the Gila River alone and at least another 750,000 acre-feet on the main stem; therefore, it was already at or above 2.8 million acre-feet per year. Since California had senior rights to 962,000 acre-feet of surplus III(b) water, there was no water left for a Central Arizona Project.

The United States and the State of Nevada both filed motions to intervene that were granted. Although it stayed neutral on the disputed issues between Arizona and California, the United States cited numerous interests, including its obligations to Mexico under the 1944 treaty, its need to protect the many Lower Basin federal projects authorized by Congress and built by the Bureau of Reclamation, and its obligations as trustee for the native tribes with water rights claims in the Lower Basin.

With the three Lower Basin States and the United States now committed to litigation, the next question was what would the Upper Basin states do? From today's perspective where the Upper Basin states actively participate in cases involving the use of Lower Basin water, it is hard to fathom that their decision was to stay out of the case.[8] This forced California to file a motion in July 1954 to join as parties Colorado, New Mexico, Utah, and Wyoming.

Again, from today's perspective, California's arguments to bring in the Upper Basin states as indispensable parties seem compelling. It made the point that the compact interpretations Arizona was seeking would impact all seven Colorado River Basin states. A good example is the dispute over how apportionments are measured. California correctly pointed out that by application of Arizona's stream depletion theory to the obligation of each basin to the Mexican treaty, Article III(c), the deficiency would be larger, thus increasing the deliveries that the four Upper Basin states would have to make at Lee Ferry for Mexico. California even quoted the comments made during the negotiations of the 1948 compact by Wyoming legal advisor W. J. Wehrli that since under III(c) Mexico's water first comes from the surplus, reducing the amount of surplus increases the amount of a deficiency, one-half of which was an obligation of the Upper Basin.[9]

The four Upper Basin states were more concerned with the possibility that participation in the case would be used against them in Congress, where they were seeking approval of the Colorado River Storage Project. "Growth and development in the Upper Basin is outrunning the water now available," Colorado's attorneys wrote in a brief to the court. "Any delay in the development of water use in the Upper Basin is a matter of very great moment to the Upper Division States. The inclusion of the Upper Division States as parties to this litigation will be urged by opponents of development in the Upper Basin, as a reason why Congress should delay the authorization of development until the conclusion of the litigation."[10] This set up the first major decision that would shape the outcome of the case.

With the United States neutral, the Upper Basin states and Arizona opposed California's motion to expand the case. Their legal logic was straightforward. The 1922 compact divided the Colorado River into two subbasins, and Arizona's suit against California was solely and uniquely a Lower Basin dispute. Their argument convinced Special Master George Haight. He denied California's motion, ruling that only Utah and New Mexico should participate to represent their Lower Basin interests. In his opinion, Haight, quoting Article I, noted that one of the primary purposes of the 1922 compact was to avoid litigation.[11] Based on his ruling, there are two possible conclusions: First Haight was focused on the legal issues in front of him and had no real understanding of the basic hydrologic facts of the case and the nuances of California's technical arguments. Second, he understood the basics of California's case for bringing in the Upper Basin states, but concluded that if this case was allowed to have a basin-wide focus, it could lead to decisions that would either seriously undermine or even void the 1922 compact—an outcome he believed would make matters on the river worse not better.[12] In his decision, he praises the 1922 compact: "This compact followed years of controversy between the states involved. It was an act seemingly based upon thorough knowledge by the negotiators. It must have been difficult of accomplishment. It was the product of real statesmanship."[13] Among the many reasons Haight gave for denying California's motion was that to join the Upper Basin states "would be a backward and retarding step with respect to the solution of problems relating to the Colorado River system."[14]

Haight unexpectedly died shortly after his critical decision and was replaced by Simon Rifkind. Rifkind ruled that the Haight decision should go to the Supreme Court for confirmation. On December 12, 1955, the Supreme Court upheld Haight's recommendation by a 5–3 vote.[15] As the case progressed under Rifkind, both combatants struggled until Arizona replaced its chief counsel with Mark Wilmer. Wilmer, using the arguments laid out by Haight as a template, changed the direction of Arizona's case to ignore virtually all of Arizona's original claims for relief.[16] Under Wilmer, Arizona amended its original bill of complaint. Arizona's rights under the 1922 compact or how apportionments were to be measured were no longer the focus. Instead, he based his argument on how Congress intended that the secretary of the interior operate Hoover Dam under the 1928 act.

The Wilmer strategy worked. After a lengthy trial in front of Special Master Simon Rifkind and a delay due to his heart attack, in 1961 he issued a special master's report that gave Arizona much of what it wanted. Rifkind's findings were challenged and resulted in a lengthy oral argument in front of the Supreme Court. Its landmark decision was published in June 1963. The decision gave Arizona the victory it needed to make the case that there was a sufficient water supply for the CAP to justify its authorization by Congress. The decision was based on how the court interpreted the intent of Congress in 1928. Under the 1928 act, Arizona was given a congressional apportionment of 2.8 million acre-feet per year of water from Lake Mead, California

4.4 million acre-feet, and Nevada 300,000 acre-feet. On all but a couple of points, the court agreed with Rifkind. One of the important provisions where the court disagreed with Rifkind was the shortage provisions. He had recommended that shortages be applied to each of the states in the same proportion as their use of Lake Mead water. The court disagreed, ruling that except for rights perfected before the 1928 act became effective, the secretary of the interior had the power to determine shortages.

It was not a perfect victory for Arizona. It did not get title to any compact water. The water supplies available to all of the contractors from Hoover Dam were subject to water availability under the 1922 compact. The questions related to Articles III(a) and (b), the basic question as to how compact apportionments should be measured and accounted for and how reservoir evaporation was to be charged under the compact, were all left unsettled. Those questions remain unresolved and matters of dispute today. Ironically, Arizona failed on all of its initial claims for relief but still won the case.

The big legal winners may have been the Native American communities and the discretion and power of the secretary of the interior. The rights of the tribes to Colorado River water were quantified and given priorities that dated back to the individual reservations. The amounts of water are significant and the priorities senior to almost all other rights on the Lower River. Along with the 1908 Winters decision, the 1963 decision is one of the two landmark cases defining the water rights of the Native American tribes in prior appropriation states. The decision was not totally disconnected from the 1922 compact. It protected the water supplies for those projects that were in place and diverting water when the 1928 act was passed. It recognized the senior status of these rights and elevated the rights of the tribes to this protected class of what we now call precompact rights.[17]

The Decision in *Arizona v. California* and the Decree

In March 1964, the Supreme Court issued the decree implementing the decision. The decree gives the secretary broad powers and discretion to manage and control the waters of the Colorado River in and below Lake Mead. The decree implements the decision giving Arizona an allocation of 2.8 million acre-feet per year, California 4.4 million acre-feet, and Nevada 300,000 acre-feet of main stem water from Lake Mead under normal water supply conditions. It provides that the secretary can make additional deliveries during surplus water supply conditions and, vital to today's conditions on the river, directs the secretary to determine shortages under low water supply conditions. A simple way to look at the decision and decree is that the court said Arizona has a right to 2.8 million acre-feet of water from the main stem but only if the water is available under the 1922 compact. If water is not available, then after meeting the needs of the present perfected rights (without regard to state lines), the secretary has the power to decide how much water the CAP and other contractors get.

Today, the 1964 decree is the primary controlling document for how Colorado River water is used and accounted for in and below Lake Mead. The 1922 compact has been given a seat on the back bench. There is still no Lower Basin compact and, with the 1964 decree and subsequent 1968 act in place, no real incentives for the states of the Lower Basin to consider one. Arizona failed to achieve its longtime goal of obtaining a ruling that the one million acre-feet of III(b) water was exclusively written to cover its uses on the Gila River. However, as a practical matter it made no difference. The river is operated as if Arizona won. Arizona has full use of the Gila River, and there is no accepted accounting for how much water is being consumptively used on the Lower Basin tributaries.[18] No one really knows the amount of beneficial consumptive use being made under the apportionments to Lower Basin under Articles III(a) and (b) of the 1922 compact.

The case has become one of the most reviewed and dissected natural resource decisions in the Supreme Court's long history. Most of the focus has been on the long January 1961 report by Special Master Simon Rifkind, the decision itself, and the dissenting opinion. The July 1955 report of the first special master, George Haight, has received far less attention. However, it was Haight's decision to reject California's motion to bring in the States of the Upper Division that set the basic direction of the case. Utah and New Mexico were made parties to the case, but only as to their Lower Basin tributaries. From today's perspective, the decision by the Upper Basin states to ally with Arizona in opposition to California's motion and stay out of the case is puzzling. However, at the time, these states considered California their enemy. It was California that opposed the ratification of the Mexican treaty. It was also California that was fighting the congressional authorization of the Colorado River Storage Project Act and the development of water projects in the Upper Basin. Further, the Upper Basin states had made their "technical agreement" with Arizona to support its stream depletion theory of accounting and had made it a part of their 1948 Upper Basin Compact.

In 1956, while the case was in progress, the States of the Upper Division successfully obtained congressional approval of the Colorado River Storage Project Act. In fact, by the time the 1964 *Arizona v. California* decree was finalized, Glen Canyon Dam, the 1956 act's marquee project, was already storing water. Would Upper Basin participation in the case have delayed its authorization and construction? No one really knows, but it is certainly a possibility. The counterargument is that under the 1922 compact, the hydrologic case for large storage above Lee Ferry is compelling, so while the participating project subsidies might have been at jeopardy, the construction of Glen Canyon Dam, itself, would have still had strong support.[19]

What their participation would have done was to focus the court's attention on the 1922 compact and its many unresolved issues. We suspect that had the focus been on the 1922 compact and facts surrounding Arizona's five claims for relief that directly involve the compact, the decision would have been mixed. Based on Rifkind's analysis of the 1922 compact, provided primarily as background, California would have likely prevailed on its diversions-less-return-flows theory of accounting.[20] Reservoir evaporation would almost certainly have been charged to the individual Lower Basin states in some equitable way. The current apportionments of main stem water, 4.4 million acre-feet for California, 2.8 million acre-feet for Arizona, and 300,000 acre-feet for Nevada would have survived as the Lower Basin's III(a) apportionments, but the outcome of the decision on the 1 million acre-feet of III(b) water is not an easy subject for speculation. A possible outcome could have been a five-way split, with New Mexico and Utah getting enough water to cover their Lower Basin tributary uses and perhaps Nevada getting a small slice, with the remainder of the III(b) water being split equally between California and Arizona. Further, with the Upper Basin states in the case, it's quite likely that the court would have interpreted Article III(c) providing guidance on the obligations of each basin to Mexico. However, speculating on the details of such a ruling is difficult. Perhaps the secretary of the interior would have again been given more discretion and power.

What seems clear to us is that had the Upper Basin states entered the case, the 1922 compact, not the 1928 act (as implemented by the decree in *Arizona v. California*) would be the primary controlling document for Lower Basin water uses. It's likely that under a Supreme Court decision that adopted the diversions-less-return-flows theory of accounting, required that Lower Basin reservoir evaporation be apportioned and charged as a compact use, and provided clarity on the Upper Basin's obligations to Mexico, there might not have been enough water to justify the construction of the Central Arizona Project at anywhere near its current capacity.[21] Would California have fared better? We don't think so. The additional III(b) water that California might have received would have unlikely been enough to even offset its share of evaporation. It might have been lucky to even have the 4.4 million acre-feet that it has today.[22]

It is clear Rifkind was depending on what we now know were bad facts in support of the decision—that he was overallocating the river's water and depending on a continuing surplus to paper over the problem. He said as much in an exchange with California attorney Northcutt Ely over Rifkind's preliminary report. Ely, during a hearing in the summer of 1960, complained that the then proposed decision placed the Metropolitan Water District, as California's junior user, at risk of shortage. Rifkind's response was that there would always be unused surplus in the system that California could depend on to keep the Metropolitan Water District's Colorado River Aqueduct full. "I am morally certain that neither in my lifetime, nor in your lifetime, nor the lifetime of your children and great-grandchildren will there be an inadequate supply of water for the Metropolitan Project," Rifkind told Ely. "I am morally certain, as certain as I am of the multiplication table, that not within the span of the ages indicated there will be any diminution either in the present uses of the Metropolitan Aqueduct or its contemplated expansion."[23]

CHAPTER 16

The Central Arizona Project

Draining the Last Drop

Forty-three years after Reclamation chief Arthur Powell Davis told the Colorado River Compact's negotiators there was plenty of water to pursue their dreams, Reclamation commissioner Floyd Dominy sat before the House Subcommittee on Irrigation and Reclamation and told a very different story.

In the wake of *Arizona v. California*, Congress in the summer of 1965 was taking up the Colorado River Basin's last grand piece—the Central Arizona Project, a canal to carry more than 1 million acre-feet a year of Colorado River water, pumped uphill, to the booming valleys of Phoenix and Tucson. The debate offered the last off ramp on the path to full development—or overdevelopment—of the Colorado River. A project to move water from the main stem inland had been Arizona's goal for almost as long as it had been a state. Now, armed with a 1940s Bureau of Reclamation engineering plan showing it was feasible and a Supreme Court decision creating a water allocation framework, it appeared the "CAP," as it was by that time called, could finally move forward.[1]

But as the region's leaders took up the task, Arthur Powell Davis's rosy 1922 optimism—enough water to meet everyone's plans, with a surplus for later—was gone. Instead, Dominy preached scarcity. "Sooner or later," he said, "and mostly sooner, the natural flows of the Colorado River will not be sufficient to meet the water demands, either in the Lower Basin or the upper basin, if these great regions of the Nation are to maintain their established economies and realize their growth potential."[2]

Neither ends nor means had changed. Basin development, backed by the financial might of the federal government, was still the plan. But the water picture had changed considerably. The hydrology could no longer be ignored. The Colorado was a 15-million-acre-foot river at Lee Ferry, not the 17.5-million-acre-foot river Delph Carpenter and his fellow compact commissioners had believed. Not only was there no surplus for the future, there was not even enough water to meet the present aspirations and entitlements of the basin states and Mexico. Thus, October 1, 1963, the date when the 1922 compact commissioners envisioned that the states would get back together to finish the job of apportioning the unallocated surplus waters of the Colorado River, came and went with no call for the commission to reconvene. Instead of additional apportionments, augmentation of the basin's water supply by importing surplus water from the Pacific Northwest became the reason to bring the basin states together.

Beyond Dominy's frank acknowledgment of a 15-million-acre-foot-per-year river, his testimony that day also illustrated a shift in the scientific approach to informing water management decisions. Gone was the attempt seen over and over, from Arthur Powell Davis and E. C. LaRue to Royce Tipton and his contemporaries, to pin down a single set of numbers that would represent a "normal" or "average" Colorado River. There was, instead, the beginning of a transition toward the use of a probabilistic framework for analyzing the river's flow and the availability of water that would result.

"There is close agreement among water experts," Dominy explained, "as to the historical water facts of the Colorado River—what the flows have been, where they have been used, and in what amounts. Such agreement, however, does not extend to projections of future conditions. To project future conditions requires the making of assumptions, which in return requires the exercise of judgment. Where judgment is involved, there is always room for honest differences of opinion. In respect to the future water supply of the Colorado, this is a classic case in point."[3]

Those judgments, Dominy explained, include questions of which part of the historical record to use in determining baseline flows in the river, the rate of future development in the Upper Colorado River Basin, the effectiveness of "water salvage," and the way in which reservoir operations are coordinated. "Differing assumptions will, of course, result in differing projections of future water supply available from the Colorado River and of its divisions between the upper and lower basins. Hundreds of water supply projection studies have been made by the Bureau of Reclamation and by the various State agencies involved. Hundreds more could be made."

"Hundreds" may have been an exaggeration, given the rudimentary state of the computer technology of the day. But the idea that analysis to support decision-making on the river could look at many scenarios, rather than looking to the Bureau of Reclamation or the U.S. Geological Survey for a single set of numbers representing *the* flow of the Colorado River, represented a revolution in the relationship between science and the river's management and development. Those who didn't like what the numbers were telling them were able to change their assumptions—about the period of record used to estimate flows, and projected future uses in different parts of the basin—to fit their needs.

Dominy and his colleagues also introduced a second important new consideration. In acknowledging that the river was headed to a state of overdevelopment, with demand for water among the basin states outstripping supply, they suggested a solution: let the development continue to expand, and make up for the Colorado River's shortcomings by augmenting it with water from somewhere else.

In introducing such an unrealistic scheme, they put the basin on the path to an ultimate reckoning: someone—whether the Upper Basin or Arizona or California or all of the above—would not get the water on which they had pinned their hopes for the future.

At the time Dominy was making this statement, the Colorado River was coming out of a twelve-year drought that was both longer in duration and drier than the 1930s. Glen Canyon Dam, the nation's second highest dam, was nearing completion. The huge reservoir behind it, Lake Powell, was slowly beginning to fill. Because filling Lake Powell would require reducing flows to downstream users, the Department of the Interior had to first referee a nasty spat between the basins over how the fill would be accomplished. It would be the first of many interbasin disputes over how the river's big reservoirs would be operated, and it set the precedent that has been used on several occasions. If the states can't or won't reach an agreement, then the secretary would have no choice but to impose their will on the basin.

Periods of Record

For the previous half century, the question of which period of record to use in analyzing the Colorado River's past flow, in order to project future water supplies, had lurked in the background, rarely discussed explicitly. Thus, for example, E. C. LaRue had argued for consideration of the droughts of the nineteenth century in understanding how much water was available for future development, and Lawrence Lawson had toyed, perhaps dishonestly, with whether and how to consider the drought of the 1930s in determining how much water was available to allocate to the Mexican treaty. But in the debates over the CAP, the question of which period of record to use became an explicit feature of the debate. During the three years Congress spent on the legislation, a number of different supply and demand scenarios were presented. Reclamation presented its view, as did a committee of engineers representing the Lower Basin states and longtime Colorado River engineer Royce Tipton on behalf of the Upper Basin states through the Upper Colorado River Compact Commission. For the long-term average natural flow of the Colorado River at Lee Ferry, Reclamation used 1906–65, which had a flow of 15.06 million acre-feet per year.[4] In doing so, they changed the period they had used in previous studies by dropping the years of 1897–1905. Had these years been included in the average the flow would have been slightly lower, at 14.9 million acre-feet per year.

Royce Tipton, working on behalf of the Upper Colorado River Commission, used a different approach. Rather than picking and arguing for a single period of record, he analyzed and presented data on a number of different periods of records: 1903–64, 1914–64, 1921–64, and 1930–64.[5]

Tipton justified the choice of 1921–64, with an average natural flow at the Lee Ferry compact measurement point of 13.95 million acre-feet per year, by pointing out that the gauge just upstream was installed in 1921. The Tipton report, prepared to consider the long-term water available to both the Upper and Lower Basins, shocked his Upper Basin clients. He concluded that if the future of the Colorado River looked like 1930–64, the amount of water available for consumptive uses in the Upper Basin could be as low as 4.6 million acre-feet per year, and the long-term deficit in the Lower Basin could be as high as 1.9 million acre-feet per year. Tipton's report also focused the basin's attention on the very real impacts of the disputed obligation of the Upper Basin to Mexico under Article III(c) of the 1922 compact. It represented a one-for-one tradeoff. If the Upper Basin's obligation to Mexico was 750,000 acre-feet per year for a total Lee Ferry obligation of 8.25 million acre-feet annually, then the Lower Basin's deficit is reduced from 1.9 million acre-feet to 1.2 million acre-feet. That 750,000 acre-feet of gain to the Lower Basin is a loss to the Upper Basin. In his report, Tipton referred to the annual delivery of 8.25 million care-feet per year as a "fictional" delivery.

To prepare for the hearings, the Lower Basin states organized a technical committee to present a common position. The committee's findings were presented at the hearing by W. Don Maughan of California.[6] The Lower Basin committee used 1896–1964 with an average annual natural flow at Lee Ferry of 14.9 million acre-feet per year. However, this committee broke new ground in how they presented and used the hydrology. The Maughan Committee suggested that the water supply of the Colorado River be presented in terms of the probability of a certain supply. For example, he testified that "probability studies indicate a 95 percent chance that the future long-term average annual runoff (at Lee Ferry) will exceed 13.3 million acre-feet, and a 50 percent chance that it will equal or exceed 14.9 million acre-feet." He went on to say "the probability approach to projecting future events is a commonly accepted technique in the appraisal of risks. . . . Many scientists and educators have encouraged the use of probability techniques in planning for water conservation." Finally, Maughan made reference to what might be the reawakening of a recognition of the value of paleohydrology. "Wide variation may be expected in the future runoff of the Colorado River not only in annual flows but also in 10-year and even 50-year averages, as demonstrated by nearly 70 years of streamflow data, about 110 years of lake level measurements and roughly 700 years of tree-ring data. For example, in the 70 year of estimated and measured flow at Lee Ferry, the average virgin flow for the first 35 years was about 17 million acre-feet annually, but the average for the last 35 years was only about 13 million acre-feet."[7]

The Maughan committee work heralded a rapid advance of the science of hydrology and its application to water supply planning and evaluation. In some respects, it was the first real scientific advancement since the publication of LaRue's Water Supply Paper 556 forty years earlier.[8]

For the Bureau of Reclamation and the boosters of the CAP, the debate over the period of record used to analyze the project had serious consequences. With a risk that Arizona's CAP rights would be junior to other users on the Lower Basin main stem,[9] a difference of a few hundred thousand acre-feet per year of available supply had a big impact at the margins. In early 1968, during the last major subcommittee hearings on the project authorization, Chairman Aspinall peppered Dominy and Udall with questions on the period of record Reclamation had chosen. Aspinall asked why start at 1906, why not 1922 (the first full year after the Lees Ferry gauge was installed) or 1914 (the starting year picked by the Upper Colorado River Compact Commission Engineering Committee) or 1897 (from the HD 419 hydrologic appendix)? Dominy responded that Reclamation was using the longest period of record for which it had confidence.[10]

It is probably not a coincidence that starting with 1906 also provided the CAP with the most water. Under the same upstream development assumptions, using the 1922–65 period instead of 1906–65 reduced the fifty-year average yield of the CAP from 1,045,000 acre-feet per year to 622,000 acre-feet per year. For the same investment, it would almost double the capital cost per unit of CAP water.[11] Thus, like Debler's decision nearly forty years earlier, Dominy's decision to pick 1906 was likely more about marketing the project than science.

Upstream Use

The differing views of the period of record to use and the resulting water supply available were only half the story of uncertainty and judgment implied by Dominy's comments. The engineering experts who testified before Aspinall's committee in 1965 also presented a variety of different estimates for upstream depletions above Lee Ferry. This mattered a great deal, because the question of how much of their 7.5 million acre-feet per year the Upper Basin's residents might use had just as much impact on the water available downstream as the natural flow of the river. While it had become clear the Upper Basin would likely never use its full 7.5 million acre-feet, there were significant differences about how much water it might eventually use. Tipton assumed that the Upper Basin states—his clients—would deplete 6.3 million acre-feet per year by the year 2000.[12] Reclamation used a year 2000 depletion of 5.43 million acre-feet per year.[13] But unlike the water availability, the Lower Basin states of Arizona and California were not aligned on Upper Basin uses. The Colorado River Board of California assumed 5.7 million acre-feet per year, Arizona assumed only 4.9 million acre-feet per year in 2000. Of course, these numbers were based on the differing interests of each state. Arizona's use of a low number improved the water supply available to its CAP. California's use of the higher number increased the shortages to existing uses on the main stem in and below Lake Mead supporting its case for a subordination of the CAP to its existing uses.[14] Once again, the players were picking and choosing the technical information needed to support their preferred political outcome.

Augmentation

The Central Arizona Project hearings may have marked the historic moment when the region's leaders acknowledged that the Colorado was a 15-million-acre-foot-per-year river, not a 17.5-million-acre-foot one. But the acknowledgment came with an extraordinary plan: to solve the problem, they would import enough water to overcome the Colorado's shortcomings. They would import 2.5 million acre-feet of water from somewhere else.

Despite their differing supply and demand assumptions, Tipton, Maughan, and Reclamation all had come to the same basic conclusion. Augmentation—importing water into the Colorado River Basin—was essential, they argued, and the legislation authorizing of the CAP needed to include an importation plan.

From the start, Dominy and his boss, Secretary of the Interior Stewart Udall, made it clear that the administration strongly supported at least comprehensive studies of importation strategies. Udall was clear: "There is one overriding fact of life which dominates the entire Colorado River Basin today—the fact that the streamflow quantities which formed the framework of the Colorado River Compact in 1922 and the Mexican treaty agreement in 1965 are not present in the river today. The specter of shortage hovers over the entire region, in 1965, and must of necessity provide the setting for all deliberations concerning its future." Udall went on to describe Interior's objectives:

> The undertaking of detailed and comprehensive studies of how and where to get additional water.
>
> Establishment of a Lower Colorado River Basin development fund to assist in meeting the cost of water development.
>
> Authorization of needed lower basin works now. These included the Central Arizona unit, Marble Canyon Dam, water salvage programs, and recreation and fish and wildlife facilities.[15]

The administration's proposed legislation included two other fundamental concepts that would become a part of the ultimate legislation. First, the Mexican treaty burden would be a national responsibility—meaning that the national taxpayers, not water users within the basin, should pay for the cost of importing the 1.5 million acre-feet of water. Second, existing uses would be fully protected—meaning that diversions by the CAP would be junior in priority to existing uses on the river including California's full 4.4 million acre-feet per year. Udall stated that the administration's bill provided that once 2.5 million acre-feet was imported into the basin, the priority given to existing users in the Lower Basin would be lifted. This 2.5 million acre-feet of imports plus the 15 million acre-feet thought available in the river just happened to equal 17.5 million acre-feet, the amount of water thought available by Delph Carpenter and fellow compact commissioners.

Compromises and a Final Bill

Pumping water uphill and over a long distance requires a lot of electricity. It is why the Metropolitan Water District signed a contract to purchase the largest allocation of power from Hoover Dam. Backers of the Central

Arizona Project needed a way to pump more water both farther and higher than MWD's Colorado River Aqueduct. They hoped to do something similar. But with the best dam sites taken, their scheme moved upstream, toward E. C. LaRue's initial vision of a staircase of dams as the river flowed downstream, wringing every last kilowatt of energy from the Colorado. Thus, the initial Central Arizona Project plans contemplated building one or two dams, both located in what is now Grand Canyon National Park—Marble Canyon Dam just upstream of the confluence of the Colorado River and the Little Colorado River in the upstream reach of the Grand Canyon, and Bridge Canyon Dam, also referred to as Hualapai Dam at Bridge Canyon, just downstream of the 1960s park boundaries.[16] Both sites had been studied and championed in the 1920s by E. C. LaRue.

But by the 1960s, the politics of water development had changed. The potential for dams in the Grand Canyon became an issue of national debate, bringing not only conservationists and recreationists but also proponents of nuclear and coal-fired electric power plants to the table.

The energy needed for the CAP as of 1965 was about 2.1 billion kilowatt-hours per year, or roughly half of what is produced annually at Hoover Dam.[17] Together Bridge Canyon and Marble Canyon Dams were designed to produce about 7.5 billion kilowatt-hours per year. For Dominy and the Reclamation planners, the additional power was important. Its sale would defray the costs of building and operating a project to import water into the Colorado River Basin. The notion of "cash register dams" was now deeply embedded in the Bureau's approach to river development, and this scheme carried it to its logical conclusion.[18]

The inclusion of dams that would bookend the Grand Canyon and help fund an importation project likely from the Columbia River Basin brought national attention. At Senate hearings in May 1967, Secretary Udall stated, "This has been one of the most controversial issues involved in Colorado River Project legislation, in fact the most controversial."[19] Udall's testimony set the table for several changes in legislative direction: First, the dams would be dropped. Instead, the Bureau of Reclamation would participate in the construction of a coal-fired plant to power the CAP's huge pumps. Second, the administration now favored legislation to expand the boundaries of the Grand Canyon National Park to include both dam and reservoir sites and forever prevent their construction. And third, it was becoming increasingly obvious that economic realities, regional opposition, and environmental concerns meant that it was highly unlikely that a large importation project would ever be built. This, paradoxically, triggered the enlargement of the CAP canal facilities so that it could divert 1.6 million acre-feet per year, not 1.2 million. This change contemplated that the CAP would pump more water in wet years to make up for the reality that with its junior priority during the inevitable drought periods like the 1930s or 1953–64, it would have little water available to pump.

Further, without augmentation, the CAP was a serious threat to the Upper Basin. Reclamation's numbers showed the average CAP supply would drop from 1.53 million acre-feet per year in 1990 to only 882,000 acre-feet per year in 2030, the result of growing water use in the Upper Basin eating into the surplus needed to keep CAP water flowing.[20] The concern was that, like California had done for decades, Arizona would oppose Upper Basin projects to preserve the supply available to the CAP.

Aspinall's bargain was to use the CAP legislation to direct the secretary to determine how much storage the Upper Basin could maintain for future drought protection and authorize five more participating projects in Colorado and an additional unit of the Central Utah Project. If indeed the CAP was going to rely on the Upper Basin's unused water then it was in the Upper Basin's interests to use its water sooner rather than later. This strategy was only partially successful. Two of the Colorado Projects, West Divide and San Miguel, were never built; the major irrigation component of the Animas-La Plata project was removed; and even the two projects that were completed, Dolores and Dallas Creek, were downsized and use less water than planned.

The Central Arizona Project
Is Finally Authorized

On September 30, 1968, the Colorado River Basin's leadership gathered in the East Room of the White House to celebrate the denouement of Arizona senator Carl Hayden's long career. "I think I know what a plentiful water supply can mean to a barren and parched countryside," President Lyndon

Johnson said as he signed the Colorado River Basin Project Act. Arizona's long dream to move water from the main stem of the Colorado River into Central Arizona was finally coming true.[21]

By the time the legislation was completed, the dream of augmentation was already fading. Rather than explicitly calling for new water to be imported to the Colorado River Basin, the act directed the secretary to study the water supply problems in the different regions of the western United States. The language making the Mexican treaty delivery a national obligation remained, but at the insistence of members of Congress from Columbia River Basin states, language was added that prevented Interior from studying importation into the Colorado River Basin for a ten-year period. "According to my reading of the law," Washington senator Henry "Scoop" Jackson told Arizona's Mo Udall, "you can't *study* it, *contemplate* it, or even *dream* about it. The only thing you are permitted to do is to *forget* about it." To which Udall responded in his autobiography, "And I almost have."[22] As a practical matter, any serious consideration of large importation projects ended with the passage of the 1968 act.

The authorization of the CAP came with strings. The most important was its junior status. If there were shortages in the future, and the operational studies made by the Bureau of Reclamation made it clear there would be, the CAP would have to absorb most of the cuts. To help in wetter years, the aqueduct was again increased to a capacity of 3,000 cubic feet per second, but use of the last 500 cubic feet per second was limited to when Lake Powell was spilling. Title V authorized the construction of the five Colorado Projects and an expansion of the Central Utah Project. Additionally, to address the nagging question of how much water the Upper Basin could hold in its big system reservoirs and not violate Article III(e) of the 1922 compact, Title VI was written to answer this seemingly simple, but actually quite complicated, question. The act sets forth the priorities for water releases from Glen Canyon Dam: first priority is to meet the Upper Basin's obligation to Mexico under the 1944 treaty, if any; second priority is the 75 million acre-feet every ten years, which is an average release of 7.5 million acre-feet per year; the third and most complicated priority is to release extra waters above and beyond the amount of water the Upper Basin needs to keep in storage to meet the first two priorities during a critical drought period without impairment to uses in the Upper Basin. If there is any extra water then it would be released so that the active storage in the two big system reservoirs, Lakes Powell and Mead, would be equal at the end the water year.[23]

Operations under the third priority is commonly referred to as "equalization." Under the 1968 act, the reservoir storage level (elevation) in Lake Powell above which these releases are made is set by the secretary after consultation with the Upper Colorado River Compact Commission and the Lower Basin states. The law instructs the secretary to consider factors such as the critical drought of record and the probability of water supply. It is one of those rare instances where an element of the law of the river directs the secretary to make a critical operational decision using the science of hydrology. Additionally, Title VI directs the secretary to coordinate the operation of the major projects authorized by the three major development acts: the 1928 Boulder Canyon Act, the 1956 Colorado River Storage Project Act, and the 1968 Colorado River Basin Project Act, and in consultation with basin states, promulgate long-range operational criteria (commonly referred to as the LROC).[24]

The Department of the Interior wasted no time in getting the states together to prepare the initial LROC. In the spring of 1969, shortly after his confirmation, President Nixon's secretary of the interior, Walter Hickel, convened a task force of state and federal representatives to prepare draft criteria. By then the Bureau of Reclamation had made significant progress in the use of the mainframe computer technology to simulate and evaluate a wide range of different operational scenarios with significant detail. It had developed the earliest generation of what is now the Colorado River Simulation System (CRSS). It was the first detailed basin-wide river model. It gave Reclamation the ability to evaluate the probability of a critical event occurring, such as Lake Powell or Lake Mead dropping below the minimum levels necessary to produce power. They now had the ability to look at flows at Lee Ferry and other locations on the river under a range of different upstream depletion and downstream demand assumptions.

What this new technology did not improve was our understanding of the hydrology of the river. The period of record used for the initial studies was 1906–66. In one way, it was a step backward—the period of 1897–1905 was not included in the simulations.[25] The advancements in modeling did nothing to help resolve the dispute within the basin over the Upper Basin's obligation to Mexico under the treaty with Mexico. In fact, the model advancements coupled with the requirements of the 1968 act provided fodder for many new disputes, including how the secretary would determine the storage level in the Upper Basin reservoirs that would trigger equalization.

At the end of October 1969, the task force and related committee issued its report.[26] In December 1969, the secretary proposed the first LROC. It provided that Glen Canyon Dam would be operated with a goal of releasing 8.23 million acre-feet per year—enough to meet the compact's 7.5-million-acre-foot-per-year average, plus one-half of the U.S. treaty obligation to Mexico, with a small adjustment for the average flow of the Paria River, which joins the Colorado River below Glen Canyon Dam but above the Lee Ferry compact point.[27] The Upper Basin states strenuously objected, primarily because of their disagreement over the way the Mexican water was included, but they ultimately accepted it—another example of the power of the secretary of the interior to impose their will on the basin states.

As the 1970s began, all of the major pieces of the law of the river were in place. The first tier are the prior appropriation doctrine, the 1922 compact, the 1928 Boulder Canyon Project Act, and the 1944 treaty with Mexico. The second tier are the 1948 Upper Basin Compact, the 1956 Colorado River Storage Project Act, the 1964 decree in *Arizona v. California*, and the 1968 Colorado River Basin Project Act. There was general agreement that the Colorado River was a 15-million-acre-foot river at Lee Ferry. Much of the water infrastructure we have today, including all of California's major projects, was in place. Plus, the Bureau of Reclamation was busy designing and building a backlog of smaller projects throughout the basin. A few of the larger municipalities were planning and building additional projects with their own resources. The basin's water managers knew that once these new projects were on line and diverting problems lay ahead. The math didn't work. Demands for the river's water would exceed supply. Fifteen million acre-feet was not enough. Despite the math problems, a culture of water entitlement still controlled the water policies of most of the states, and few politicians or water managers even dared suggest that the era of river development was nearing an end.

As luck would have it, nature again helped delay the inevitable. Conditions from the early 1970s through the late 1990s would again be relatively wet. These wet conditions, the slow pace of construction on major projects such as the CAP, and the cancelation of a number of Upper Basin projects temporarily covered up the math problem. This period of false security would come crashing to an end in 2002.

CHAPTER 17

Tree Rings and Climate Change

In the years that followed approval of the Central Arizona Project, two new kinds of science emerged that changed the way the river management community thought about Colorado River hydrology, past and future. The use of tree rings to estimate past climates allowed scientists to reach back far beyond the river's record of gauged flows, providing a much more complete picture of the Colorado's range of variability. And the growing understanding of the effect of rising greenhouse gases suggested a future of declining flows. Importantly, both these new approaches to the Colorado's hydrology were being developed by scientists outside of the traditional water management agencies that had long controlled the Colorado River's scientific discourse. And both were initially poorly received.

Tree Rings

At the December 1977 meeting of the Colorado River Water Users Association, the Bureau of Reclamation called on Mike Clinton and William Lane to talk about a paper on the water supply of the Colorado River recently published by two tree ring scientists, Charles Stockton and Gordon Jacoby. Stockton and Jacoby had concluded that long-term average natural flow at Lee Ferry was 13.5 million acre-feet per year, 10 percent below the then accepted value of about 15 million acre-feet per year. Since Dominy had testified a decade earlier that even with 15 million acre-feet at Lee Ferry there was not nearly enough water to meet future needs, the idea that there were only 13.5 million acre-feet was troubling. It would mean many of the projects then under construction or in the queue would not have a reliable water supply.

The 1970s was a difficult time for the Bureau of Reclamation. The decade began with a broad challenge to the value of reclamation projects.[1] Virtually all of its ongoing construction projects were far over budget. Progress on the Central Arizona Project was bogged down by squabbles over Arizona's groundwater management and the location of the project's storage reservoir near Phoenix. In 1976, Teton Dam, a Reclamation project in Idaho, catastrophically failed, casting serious doubt on the agency's engineering and management skills. In 1977, the Carter administration identified a list of Reclamation projects it believed should not be built or even completed. Finally, the agency's name had been changed from the Bureau of Reclamation to the Water and Power Resources Service. Dropping the term *reclamation* was a sign that the agency's core purpose of reclaiming arid lands through water development to encourage the settlement of the west was no longer relevant to a region where growth was exploding due to quality of life and economic opportunities in the West's thriving urban centers.

Thus, Lane and Clinton had the task of not only addressing the Stockton and Jacoby paper. They also needed to restore some confidence in their ailing agency. They devised a clever response. They told the audience their agency considered the Stockton and Jacoby report to be a significant scientific advancement, but there were still too many problems for the results to be used for management or decision-making purposes. They recommended supporting further research.[2]

Use of tree rings to evaluate droughts and estimate river flows in the distant past was developed well before the 1970s. A. E. Douglas, an astronomer, first recognized in the early years of the twentieth century that Arizona's ponderosa pine tree rings recorded the variation in precipitation from one year to the next.[3] His graduate student, Edmund Schulman (1908–58), was the first to use moisture-sensitive trees as a proxy for stream flow as well as precipitation. Schulman published two papers in the 1940s: the first, in 1942,

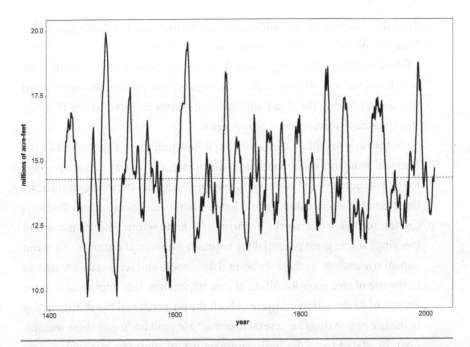

Figure 5 Tree ring reconstruction of Colorado River flow, ten-year moving average. *Source:* "Colorado River at Lees Ferry, CO," TreeFlow, https://www.treeflow.info/content/upper-colorado.

"A Tree-Ring History of Run-off on the Colorado River," was sponsored by the Los Angeles Bureau of Power and Light (a major user of Hoover Dam Power); and the second, in 1945, "Tree-Ring Hydrology of the Colorado River Basin," was based on his PhD dissertation.[4]

Although Schulman's reconstructions were limited by the lack of computers and less-sophisticated statistical methods, he came to several conclusions important for Colorado River decision-makers. First, the period from the early 1900s to about 1930 was unusually wet. Second, the period from about 1870 to 1900 was dry. Schulman thus confirmed the findings of LaRue, Stabler, and Sibert that a multidecadal dry period preceded the early 1900s. His work further suggested that there were many such long-term dry periods in the basin's long history.

There is no question that the Colorado River water community was aware of Schulman's work. During the negotiations of the 1948 Upper Colorado

River Basin Compact, negotiators made several mentions of extended droughts from the past shown by tree ring analysis. The tree ring research was also referred to during the congressional testimony for both the 1956 and 1968 acts. However, as noted by Reclamation's Lane, "progress of this application was slowed during the 1950's and 1960's. It is only in recent times that that tree rings have received new prominence."[5]

In the early seventies, the Lake Powell Research project sponsored by the National Academy of Sciences funded Stockton and Jacoby to use tree rings to reconstruct flows at Lee Ferry. In response to a draft they were circulating, the Bureau of Reclamation tasked Lane, from its Engineering and Research Center, to review the work. In early 1976, Lane wrote, "While the use of tree rings shows great potential for improving historical estimates for mean annual streamflow . . . there are several drawbacks and serious shortcomings in the use of tree rings for filling in past streamflow. It is important that the Bureau of Reclamation be aware of both the potential and the shortcomings in the use of tree rings for several reasons."[6] He went on to give three reasons. First, he stated that "the indiscriminate use of synthetic streamflow data generated through the use of a mathematical relationship utilizing tree-ring data may be highly unwarranted and unjustified." Second, Lane made the argument that even a three-hundred-year reconstruction of filled-in data, if the mathematical relationship is weak, "may be of no more statistical value than five years of actual data." Finally he addressed the political concern head on: "The recent public announcements of a new estimate based on tree rings for the mean annual virgin flow at Lee Ferry, Arizona, of 13.5 million acrefeet . . . is considerably lower than previous estimates . . . [and] will attract a great deal of attention."

Stockton and Jacoby's signature paper "Long-Term Surface-Water Supply and Streamflow Trends in the Upper Colorado River Basin" was formally published in March 1976.[7] The report went beyond a scientific paper on the reconstruction of the flows based on dendrochronology. With boldness that echoed E. C. LaRue, it drew a number of conclusions concerning the adequacy of the water supply of the Colorado River Basin. The primary conclusion was that the long-term average natural flow of the Colorado River at Lee Ferry is 13.5 million acre-feet plus or minus 1 million acre-feet. Stockton and Jacoby noted, "Historically, the Upper Basin has not utilized its entire apportionment and much of its allowable depletion has passed to the Lower Basin. This has created an inaccurate sense of surplus in some minds, but closer examination based on tree-ring analyses shows that the UCRB is in fact headed for a water shortage." They also concluded that "the period of 1906 through 1930 was the greatest extended period of high surface runoff from the UCRB within the last 450 years. Consequently, any estimates of future flow that are based on periods of record which include this wet interval tend to be inflated." Subsequent tree-ring-based reconstructions have confirmed Stockton and Jacoby's conclusion concerning the unusual wetness of 1906–30; however, the Bureau of Reclamation continues to include 1906–30 in the period of record used for most planning purposes.[8]

While the Bureau of Reclamation was slow to endorse and use tree-ring-based reconstructions of stream flows in the Colorado River Basin, the methodology gained acceptance throughout the basin water community and in other watersheds throughout the West. By the 1990s, major water agencies, such as Denver Water and the Salt River Project, were funding flow reconstructions on streams that they were utilizing for water supplies. Even though water supply conditions from the 1970s through the 1990s were generally wet, it was a sign that water providers were taking seriously the message from the tree-ring-based reconstructions that past droughts on the Colorado River have been longer in duration and more severe than those that were during the 1900s.

Stockton and Jacoby's 1976 paper and subsequent reconstructions all show a number of significant droughts in the river's past. Fascination with the extreme droughts in the tree ring record led the basin's academic community to examine how the basin would handle a severe and sustained drought. The results were published in the American Water Resources Association's October 1995 Water Resources Bulletin.[9]

The study focused on a drought from 1579 to 1600, a twenty-two-year period with a reconstructed average natural flow at lee ferry of 11.05 million acre-feet per year. This simulated drought was severe enough to drain Lake Powell, requiring the Upper Basin states to curtail existing uses to meet 1922 compact obligations and causing shortages in the Lower Basin. The study evaluated a range of economic, legal, environmental, institutional, and social issues. Unfortunately, the study results were neither broadly discussed nor considered relevant by the states and water agencies. This may have been because, when the study was published, the reservoirs were full. But water managers also saw the study as a broad attack on the Law of the River. The study's primary recommendation was that the basin states and federal government "explore the possibility of replacing the 1922 Compact" with a more modern version that among other things mandated the consideration of meeting nonconsumptive water demands and that established long-term allocations of water in proportion to current demands, rather than the 1922 compact apportionments.[10]

While the severe and sustained drought study failed to influence water management in the basin, its use of tree-ring-based reconstructed flows was a breakthrough. By the early 2000s, the long-term flow record had been reconstructed back to AD 762, and the use of tree-ring-based reconstructed flows for evaluating drought management strategies had gained acceptance as a planning tool. The latest tree-ring-based long-term (1416–2018) average natural flow of the Colorado River at Lee Ferry is 14.31 million acre-feet per year—very close to the Sibert board's recommendation in 1928.[11]

Returning "Reclamation"

As the calendar turned to 1981 and Ronald Reagan was sworn in as the nation's fortieth president, it brought with it a momentary period of a bright future for the beleaguered Water and Power Resources Service. One of the new administration's first actions was to restore its name to the "Bureau of Reclamation." The Colorado River water agencies and their boosters were full of hope that federal money would again flow to build their long-stalled projects. That hope was short lived. Reagan's Interior appointees suggested a path forward based on the concept of "cost sharing." The projects on Carter's "hit list" would be resurrected, and the federal government would be happy to build them with the states' help—in the form of a 40 percent state share of the capital costs. The joke within the water community was that while Carter bludgeoned the projects, Reagan loved them to death, but in both cases they died.

By the early 1990s, except for the completion of a few projects serving Indian communities, almost all of the Colorado River Basin projects that are likely to be built had been completed.[12] Beginning in the late 1970s, much of the Bureau of Reclamation's management attention focused on the implementation of federal environmental statutes passed during the late Johnson and Nixon administrations. The Colorado River Basin Salinity Control Act, the National Environmental Policy Act, the Clean Water Act, and the Endangered Species Act were now in the forefront of design and operation of existing and under-construction projects. As an agency, the Bureau of Reclamation had been transformed from one that built its reputation on engineering and construction to one struggling to incorporate the values of recreation, water quality, and the environment into its decisions.

As luck would have it, the 1980s brought very high flows. After a dry start in 1981, 1982–87 were above average and 1983–86 were big years. In 1983, high flows that surprised forecasters filled Lake Powell to the brim, causing big spills that threatened to undermine the integrity of the Glen Canyon Dam's emergency spillways. After a drought period from 1988 to 1994 that mostly went unnoticed by the Colorado River Basin (except in California), high flows returned through the late 1990s.

The Emergence of Climate Change

Called on by a *New York Times* reporter in October 1983 to explain the implications of the nascent science of the impacts of rising greenhouse gases, Paul Waggoner explained that water in the West would become too precious to waste on crops as climate change sapped flows in the Colorado River.[13] Waggoner and his colleague Roger Revelle had concluded that a two-degree-Celsius increase in temperature (a little less than four degrees Fahrenheit) would reduce the Colorado River's flow by nearly 4 million acre-feet per year at Lees Ferry.[14]

Their work came as scientists were first coming to terms with the implications of rising greenhouse gases in earth's atmosphere. It was becoming clear in the 1970s and early 1980s that carbon dioxide from the burning of fossil fuels was ever so slightly reducing outgoing radiation, jiggering earth's energy balance and raising questions about the impact on things like water supply.

The first attempt to sort out the impacts of climate change on water supplies had come four years earlier, by one of the pioneering tree ring researchers, Charles Stockton. Working with colleague William Boggess, Stockton had written a major report for the U.S. Army Corps of Engineers engaging two "what-if" questions. Given that climate and therefore river flow was never stable on multiyear time scales—an insight that can be traced to LaRue and that showed up clearly in Stockton's tree ring work—what would the implications of a colder climate be on water supplies? What of a warmer one?[15]

Stockton and Boggess were not explicitly predicting warming but rather generating early data to help decision-makers prepare for a range of possibilities. At the time, climate research in general was pointing toward warming as a result of rising greenhouse gases, but there was still some uncertainty, and a minority of scientists were still considering the possibility of an offsetting cooling trend.[16] On the "if it got warmer" side, Stockton and Boggess predicted there would be less water.

It was Revelle and Waggoner who first argued persuasively which side of that "what-if" dilemma the Colorado River Basin faced. While it would be some time before the effects would be felt, the National Research Council report of which their work was a part was an effort to get ahead of the problem. "We believe there is reason for caution, not panic," William Nirenberg, chairman of the assessment committee that produced the report wrote in its introduction. "Since understanding and proof of what is happening to climate as a result of practices that load the atmosphere with CO_2 may come too late to allow for corrective action, we may not be able to wait to make certain there is a best course."[17]

On the scientific side, the decades that followed offered an increasingly insistent drumbeat of research regarding Colorado River flows refining the findings of Stockton, Boggess, Revelle, and Waggoner. "Even in the absence of changes in precipitation patterns, higher temperatures resulting from increased greenhouse gas concentrations lead to higher evaporation rates, reductions in streamflow, and increased frequency of droughts," a 2000 assessment for the U.S. Global Change Research Program concluded.[18]

By 2007, a group convened by the National Academy of Sciences summed up the status of the science in this way: "A future in which warmer conditions across the region are likely to contribute to reductions in snowpack, an earlier peak in spring snowmelt, higher rates of evapotranspiration, reduced late spring and summer flows, and reductions in annual runoff and streamflow."[19]

"Any future decreases in Colorado River streamflow, driven primarily by increasing temperatures, would be especially troubling because the quantity of water allocations under the Law of the River already exceeds the amount of mean annual Colorado River flows," the report went on. "This situation will become even more serious if there are sustained decreases in mean Colorado River flows. Results from these numerous hydroclimatic studies are not unanimous, and all projections of future conditions contain some degree of uncertainty. Nevertheless, the body of climate and hydrologic modeling exercises for the Colorado River basin points to a warmer future with reductions in streamflow and runoff."[20]

Yet just as Colorado River Basin water management was slow to respond to the warnings of LaRue, Stabler, and Sibert, the water managers of the late twentieth and early twenty-first century were slow to acknowledge the risks posed by climate change and slower still to incorporate them in the hydrologic analyses used to support major decisions.

When the Bureau of Reclamation completed a major analysis of river flows in 2000, it offered a response straight out of the playbook used in reaction to the Stockton and Jacoby tree ring research two decades earlier. There was too much uncertainty about the impact of climate change to incorporate it into the hydrology analyses being used to support development of the Interim Surplus Guidelines then under development, the Bureau concluded. At the time, there had been nearly a dozen scientific studies looking at the implications. While generally pointing toward a risk of decreasing flows, the studies still wrestled with significant uncertainty.[21] That was the peg on which the Bureau of Reclamation hung its planning hat. "There is not yet a consensus in the scientific community regarding whether long-term climate change will result in overall wetting or drying of the Colorado River Basin," the Bureau reported in the Final Environmental Impact Statement for the 2001 Interim Surplus Criteria.[22]

Six years later, as the Bureau engaged in another major analysis of the risks of shortage on the Colorado River, the agency took a major step toward recognizing the importance of climate change. By this time, analyses done under the auspices of the National Environmental Policy Act had become, for the Bureau of Reclamation, the primary platform for technical analysis of the river's hydrology to inform major decisions—the forum for the sort of work once done by LaRue, Stabler, Sibert, Debler, and the Fall-Davis report. Reclamation convened a Climate Change Technical Work Group of independent experts to review the state of the science and offer recommendations for its inclusion in river management decision-making processes. Their findings, published in 2007, echoed the similar work published that year by the National Academy panel: "Although an aggregate message from these studies may be that a decrease in runoff can be expected, runoff response across these same studies *ranges from increase to decrease.* . . . The degree to which current methods can provide reliable information about future streamflow variability remains a question" (emphasis in the original).[23]

With that hesitancy from the Bureau's panel of experts, the analyses done for the report, on which decisions about how to share shortages as reservoirs declined, followed the traditional approach, using recorded streamflow beginning in 1906. That meant that not only did climate change remain off the table for planning purposes but so did the droughts of the late 1800s identified nearly a century before by LaRue, Stabler, and Sibert. The resulting decision framework again offered a rosy scenario regarding how much water was available for the basin's water users—a scenario the faults of which would quickly emerge.

CHAPTER 18

Reality Sets In

In the summer of 2000, the Lower Basin states were taking water out of Lake Mead like there was no tomorrow. Full nearly to its spillways at 1,214 feet above sea level in January 2000, Mead offered a surplus, or at least the illusion of a surplus, the mirage on which the house of cards of Colorado River water allocation had been built.[1] With the Central Arizona Project built out, Arizona was finally able to take its full 2.8 million acre-feet of water from the Colorado River's main stem. Nevada consumed 322,000 acre-feet, more than its legal allotment, and California took nearly 5.2 million acre-feet, far more than the 4.4 million acre-feet allocated by the Boulder Canyon Project Act. Even Mexico got into the act, with a 2 million acre-feet delivery in 2000, half a million acre-feet above its normal treaty share.[2] Colorado's Front Range, Utah's Wasatch Front, Las Vegas, Phoenix and Tucson, and the Sothern California coastal plain were growing rapidly, but the wetness obscured growth's hydrologic cost.

But as Lake Mead dropped—nearly twenty feet in 2000, and another nineteen feet the following year as Nevada, Arizona, California, and Mexico continued to take more water out than flowed in—it was finally becoming clear that "tomorrow," the time of reckoning with the Colorado River's over-allocation, had arrived.

The next decade would see a revolution in our understanding and management of the Colorado River. The combined effects of overdevelopment of the Lower Basin's available water supplies, deep drought, and the impact of a warming climate caused reservoir storage to drop from nearly full to unprecedented lows.[3] There was now no way to avoid the conclusion that the Colorado River was overtapped and headed for trouble.

In response, two important and seemingly conflicting stories emerged. The first is that the crisis brought to the surface the cumulative impacts of past decisions that had been justified based on a flawed understanding of the hydrology of the Colorado River. The second is that this same crisis opened the door for new approaches to managing the resources of the Colorado River, better and more effective communications among the basin's stakeholders, and cooperative efforts that prior to the twenty-first century might have seemed unimaginable.

The Completion of the CAP and the California 4.4 Plan

With hindsight, it is possible to date the beginning of the twenty-first-century changes on the Colorado River to March of 1985, when Arizona pumped the first water from Lake Havasu into the Central Arizona Project canal. "Now as a trickle but soon as a torrent," Bill Boyarsky wrote from Phoenix in the *Los Angeles Times*, "Arizona is finally taking its share of the Colorado River, and the impact will be felt from here to the Pacific beaches."[4] Boyarsky's warning was explicit—the surplus water that until then had flowed west to Southern California would now be headed east, toward the Phoenix metro area, which was growing at a rate of more than 75,000 people a year.[5]

After authorization in 1968, CAP construction began in the early 1970s, but the project took over two decades to complete. Pumping steadily increased through the 1990s, first topping 1 million acre-feet in 1993. By the late 1990s, with the project complete, Arizona was able to take its full 2.8 million acre-feet of water per year from the main stem of the Colorado. As soon as it could, it did. If the demands were not sufficient, Arizona pumped it from the river to the central part of the state, storing it underground in the deserts near Phoenix.[6]

With Arizona now taking 2.8 million acre-feet per year and with explosive Las Vegas growth causing Nevada to fully use its 300,000-acre-foot main stem apportionment, the long delayed day of reckoning for California was in sight. California was now facing the problem it had feared for decades: a real water limit of 4.4 million acre-feet per year. Avoiding this was why California refused to accept a 1930 Lower Basin compact that would have given it a 500,000-acre-foot main stem apportionment of III(b) water—the extra million feet per year of Lower Basin water added to the compact at the last minute to placate Arizona. It is why California fought ratification of the 1944 treaty with Mexico and the passage of the 1956 Colorado River Storage Project Act. Avoiding this day was the primary goal of California's defense in *Arizona v. California*, and preparing for this day was the reason it used its political muscle to force Arizona to accept a priority for the CAP junior to all of California's 4.4 million acre-feet.

In the late 1990s, even as the reservoirs were full, basin leaders began to express concern with California's continuing use of up to 5.3 million acre-feet per year, 900,000 acre-feet more than its 4.4-million-acre-foot apportionment. Now that the CAP was online, the math would no longer work. Since the closing of Glen Canyon Dam in the 1960s, inflow to Lake Mead had averaged about 10.7 million acre-feet per year. With California taking about 5.2 million acre-feet, Arizona 2.8 million acre-feet, Nevada 300,000 acre-feet, and Mexico 1.5 million acre-feet, and evaporation and system losses taking another 1.4 million acre-feet per year, the total demand on Lake Mead water was about 11.2 million acre-feet per year.[7] The deficit was about 500,000 acre-feet per year. This deficit had been masked by the slow buildup of CAP diversions and wet conditions. From 1980 to 1999, natural flow at Lee Ferry had averaged 16.1 million acre-feet per year, 107 percent of the long-term average. It was the wettest two decades since 1930.[8]

The solution was to use the discretion given the secretary under the 1964 *Arizona v. California* decree to determine when and how much surplus water is available annually from Lake Mead and prepare formal surplus guidelines. Because Lake Mead was then full, under average hydrologic conditions, with a natural flow at Lee Ferry of about 15 million acre-feet per year the 500,000-acre-foot deficit would cause Lake Mead to slowly fall over a decade or more before reservoir storage would reach unacceptably low levels. Under the plan, California would ramp down its demands. The surplus needed by California would slowly shrink, giving California time to phase in the conservation needed to reduce demand to 4.4 million acre-feet per year. The intent was to give California a soft landing.

Since the Sibert board had first warned about overallocation in the 1920s, it had been the subject of fierce political and legal battles, but slower development in the Upper Basin and the well-timed wet years of the 1980s and 1990s meant that the need for real on-the-ground solutions had been avoided. For California, the path to the soft landing had always been obvious but politically and economically painful. Farmers in the Imperial Valley and the other smaller desert agricultural water districts had rights to the first 3.85 million of California's 4.4-million-acre-foot apportionment, leaving the Metropolitan Water District's 1.2-million-acre-foot Colorado River aqueduct with only a 550,000-acre-foot firm supply.[9] As the supplier of municipal water to nearly twenty million Californians living in the Southern California Coastal Plain, MWD needed more than 550,000 acre-feet. To add to its supply, MWD needed to negotiate, fund, and implement measures to transfer water from the senior 3.85 million acre-feet of agricultural water to its aqueduct.

In 2001, after holding a series of discussions with the basin states, the secretary of the interior issued formal surplus guidelines.[10] Under the assumption that the 1906–97 hydrology, an average natural flow of about 15 million acre-feet per year, would continue into the future, the statistics suggested that while shortages would occur in the future, there would be time to implement a soft landing. Nature, as it turned out, had other plans.

The Drought of 2000–2004

With natural flow at Lee Ferry of 21.2 million acre-feet per year, 43 percent above average, 1997 was a big year on the Colorado, while 1998 was 15 percent above average and 1999 was 11 percent above average. Then conditions dried out, and the Colorado River Basin plunged into the deepest five-year drought in its post-1900 history. From 2000 to 2004, the average natural flow at Lee Ferry was 9.46 million acre-feet per year.[11] With a Lee Ferry natural flow of 5.87 million acre-feet, 2002 was the driest individual year of the drought and the second driest in the modern record, behind only 1977.

Reservoirs, full in late 1999, plunged. Lake Powell was hit especially hard. In early April 2005, active storage in Lake Powell reached its postfilling low of 7,956,600 acre-feet (elevation 3,555 feet above sea level). Being downstream of Lake Powell, Lake Mead's inflow was dictated by reservoir operating rules, not nature's supply of water. In the early drought years, the "minimum objective release" of 8.23 million acre-feet per year cushioned Lake Mead and therefore the supplies available in the Lower Basin. Although Lake Mead was slower to respond, Lower Basin water managers fully understood that the drought would eventually catch up with them. With only 8.23 million acre-feet from Powell, plus a little inflow between Lee Ferry and Lake Mead, the reservoir was operating under deficit conditions, about 2 million acre-feet per year in 2002. A deep drought meant those deficit conditions could last for many years. This was the "structural deficit"—the exact situation Northcutt Ely and Royce Tipton had warned the basin about in previous decades.

The solution was to reduce the deficit as much as possible. California was the first to be affected. In 2002, it diverted 5.28 million acre-feet. In 2003, Interior secretary Gail Norton cut California back to 4.4 million. The ramp down to 4.4 million acre-feet envisioned by the 2001 surplus guidelines became a cliff. California's long history of diverting more than 4.4 million acre-feet per year was over. However, even after cutting California to 4.4 million acre-feet, there was still a deficit, now about 1.2 million acre-feet per year.

With the 2001 surplus guidelines now moot, water agencies struggled. The problem was no longer how to divide a surplus. The problem was what would happen if the dry hydrology continued. The basin needed shortage criteria. In the spring of 2005, the Upper Basin states requested that the secretary of the interior consider releasing less than the 8.23 million acre-feet per year minimum objective release. The letter emphasized that the minimum objective release was a goal, not a guarantee. That meant the secretary of the interior had the discretion to change it. The states' letter pointed out that the drought had hammered Lake Powell much more than Lake Mead and that water supplies in the Lower Basin were abundant compared to those in the Upper Basin.[12] Therefore, the Upper Basin states believed it was appropriate to ask the secretary to reduce releases to 7.5 million acre-feet for water year 2005.

In a carefully crafted response, Secretary Norton wrote that while she agreed the secretary had the authority to reduce releases, she would not do so in 2005. It helped that the winter of 2005 was wetter than average and that forecasted inflows would increase the water stored in Lake Powell.[13] The secretary's response included direction to the Bureau of Reclamation to begin consultation with the basin states for Lower Basin shortage criteria and a more coordinated operation of Lakes Mead and Powell. Norton's message was clear—if the states could not find a consensus solution among themselves, she would impose a solution on the basin.

As the negotiations among the basin states began, the individual states took positions that were as much about the past as the future. Their internal debate played out in a manner that would not have surprised Aspinall, or Tipton, or Ely. Despite the CAP being legislatively designed to have a diminishing water supply, Arizona's goal was to protect the project's full capacity, about 1.5 million acre-feet per year. Its strategy was to transfer as much risk as possible to water users in other states. It proposed small shortages that would not be triggered until Lake Mead had very little storage. Now that California was down to its basic 4.4 million acre-feet allocation, California wanted to protect the seniority of that 4.4 and avoid the big shortages that might cut into it. California proposed larger shortages that would be implemented higher in the reservoir. Nevada needed assurances that Lake Mead would not drop below a level that would leave Las Vegas pumps high and dry before they could complete their new intake and pumps that tapped into the reservoir at the very bottom.[14]

The Upper Basin states had a number of goals. They wanted to avoid litigation, while preserving their legal arguments concerning the Mexican treaty. Since their total use, about 4.5 million acre-feet per year, was far below their 1922 compact apportionment of 7.5 million acre-feet, they wanted to preserve the opportunity for future development. They wanted to keep as much water as possible in Lake Powell. Finally, the state officials wanted to avoid being accused of solving the Lower Basin's overuse problems at the expense of their present and future water users.

Negotiations proceeded quickly for a water matter. In addition to determining how shortages would be administered by the secretary, the effort produced a more sophisticated and coordinated operation of Lake Powell and Lake Mead. The hydrologic modeling conducted by the Bureau of Reclamation to support both the negotiations among the states and the environmental analysis was sophisticated and advanced. By then, the Bureau of Reclamation had made major improvements to its river model, the Colorado River Simulation System model, or CRSS. CRSS could be used to evaluate a number of different scenarios under a range of hydrologic assumptions. In addition to using the period of record from the natural flow data base (1906–2002), alternatives were evaluated using reconstructed flows developed by the tree ring scientists.[15]

Although using the expanded hydrologic record from the tree-ring-based reconstructions was a step forward, modeling results based on the 1906–2002 record dominated the discussions among the Bureau of Reclamation and the basin states. By 2007, the climate science was becoming clear and there were numerous studies that warned of continued declines in the flows at Lee Ferry because of climate change. Like the science provided by Stabler and LaRue in the 1920s, eighty years later, the decision-makers largely ignored the inconvenient science.

In 2007, Secretary of the Interior Dirk Kempthorne signed a record of decision implementing the *Colorado River Guidelines for Lower Basin Shortages and the Coordinated Operations of Lake Mead and Lake Powell*, commonly referred to as the 2007 Interim Guidelines. The new guidelines incorporated and superseded the 2001 Interim Surplus Guidelines and were designed to satisfy the 1968 act LROC requirement.

The 2007 Interim Guidelines represent a bridge between the river's past, complete with embedded faults that led to overallocation its waters, and its uncertain future. The 2007 Interim Guidelines were designed to be temporary. The theory was that the basin needed to try them out, see how they operate, learn from their implementation, then come back in 2020 to begin negotiating new guidelines that would be put in place beginning with water year 2027.

Lower Basin shortages were tied to the storage elevation in Lake Mead. The lower the level of Lake Mead storage, the higher the level of shortage.

There are three shortage tiers—400,000 acre-feet of cuts below elevation 1,075 feet; 500,000 acre-feet below elevation 1,050 feet; and 600,000 acre-feet below elevation 1,025 feet. It was assumed Mexico would share in the shortages, something codified later in supplemental agreements made under the U.S.-Mexican treaty known as Minute 319 and its successor Minute 323. True to the language of the 1968 act, Arizona's Central Arizona Project takes the brunt of the U.S. shortages—91 percent. Southern Nevada's water supply to the Las Vegas area takes a small shortage—9 percent—and California takes no shortages to its 4.4-million-acre-foot basic apportionment. Everyone understood that a 600,000-acre-foot shortage only covered half of the projected 1.2-million-acre-foot structural deficit, but half a loaf was viewed as better than no loaf at all.[16]

The new coordinated operations of Lakes Mead and Powell were an improvement on, but not a break from, the operation of the system under the 1970 LROC. In theory, the two major elements of the original LROC were preserved, a minimum objective release of 8.23 million acre-feet per year and delivering excess Upper Basin water downstream through equalization. However, through the use of storage based triggers in Lake Powell linked to the actual storage in Lake Mead, the 2007 Interim Guidelines were designed to be more flexible and to more successfully balance storage in Lakes Mead and Powell through both droughts and recovery periods.[17] The design of the 2007 Interim Guidelines is that a release of 8.23 million acre-feet per year is the fulcrum, but when storage conditions in Powell were better than Lake Mead, but not high enough for equalization, annual releases could be increased up to 9 million acre-feet per year, to better balance storage. And likewise, if storage conditions in Lake Mead were better than Lake Powell, then annual releases could be reduced to 7.48 million acre-feet per year. In theory, and based on the hydrology modeling assumptions that were used for the 2007 Interim Guidelines, the frequency of 9-million-acre-foot releases would be about the same as the 7.48s, preserving the status quo of the 8.23.[18]

Although the term *continuing drought* remained a talking point with most of the basin states, the severe drought probably ended with water year 2005, which had a natural flow at Lee Ferry of 17.1 million acre-feet (115 percent). From 2005 to 2010, inflows to Lake Powell were about average, sufficient to maintain storage but not good enough to recover basin storage. The 2011 wet year was followed by two more drought years, 2012 and 2013, causing storage in Lake Powell to again plunge, raising widespread concern among the basin's water managers that the hydrology underpinning the 2007 Interim Guidelines was too optimistic, and shortages well beyond those required by the guidelines were likely to be needed.

At a meeting among the states and Secretary Jewell in 2013, the states and the secretary agreed that it was time to go beyond the 2007 Interim Guidelines and began preparing drought contingency plans. In the Lower Basin, this plan was necessary to finish the job the 2007 Interim Guidelines began but this time not stopping halfway there. Although it was caused by hydrology, not depletions in the Upper Basin, Royce Tipton's predicted Lower Basin structural deficit of about 1.2 million acre-feet per year had arrived, and it had real consequences. Shorting users in the Lower Basin by 600,000 acre-feet does not cover a 1.2-million-acre-foot deficit.

The effort to come to terms with the failure to heed LaRue, Stabler, and Sibert was underway, but the job was only beginning.

CHAPTER 19

CHAPTER 19
A Bridge to an Uncertain Future

By the beginning of the second decade of the twenty-first century, the problems left by water allocation policies that failed to heed the scientific cautions of LaRue, Stabler, and Sibert had taken a very specific form.

California, after decades of dependence on "surplus" water that was not really surplus, had already cut its use of Colorado River water from more than 5 million acre-feet per year to a steady 4.4 million acre-feet per year. Nevada's Las Vegas area, after exceeding its 300,000-acre-foot allotment in the early 2000s, had successfully conserved its way to less than 250,000 acre-feet per year. Arizona, in defiance of the warnings when the Central Arizona Project was being authorized in the 1960s, still clung to its 2.8 million acre-feet per year of Colorado River water use. And Lake Mead kept dropping.

Upstream, the Upper Basin was only using 4–4.5 million acre-feet in a typical water year, far below the 7.5 million acre-feet promised by Arthur Powell Davis, Delph Carpenter, Herbert Hoover, and their Colorado River Compact Commission colleagues. Yet total storage in the river's two great reservoirs, Lake Mead and Lake Powell, kept dropping.

Everyone could reasonably argue they were living within the rules. But rules cannot conjure water, and it was clear the rules needed to change.

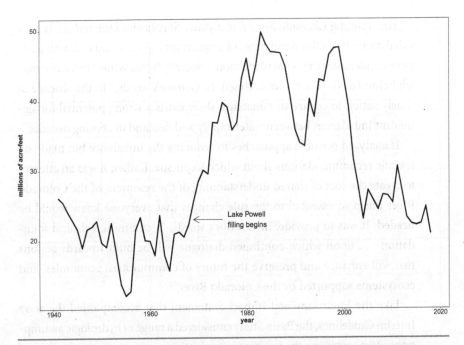

Figure 6 Combined Storage, Lakes Mead and Powell. Source: U.S. Bureau of Reclamation.

The 2012 Colorado River Basin Study

When Mike Connor took over as commissioner of Reclamation in 2009, he was mindful of the legacy he was inheriting, a lineage linking him to people like Arthur Powell Davis and Floyd Dominy. But he brought a different sort of expertise to the task—"a water lawyer skilled at making collaborative deals rather than an engineer intent on building his way out of problems."[1]

Seeing science as a path to dealing with the basin's problems, he used a legislative tool he had helped build—the SECURE Water Act, a law he wrote as a Senate aide prior to joining Reclamation. With money provided by the law, Connor and his agency undertook the most ambitious attempt at quantifying the Colorado River's supplies, along with current and future demands, that had ever been attempted. It was an attempt to do what had been promised by Interior secretary Ray Lyman Wilbur at the 1930 groundbreaking for the construction of Hoover Dam—to use scientific expertise to solve problems.

The resulting *Colorado River Basin Water Supply and Demand Study* provided more than a thousand pages of maps, charts, graphs, and technical analysis to underscore the point made more viscerally by the white "bathtub rings" left behind as the reservoirs declined. In Connor's words, "In the absence of timely action to ensure sustainability, there exists a strong potential for significant imbalances between water supply and demand in coming decades."

It analyzed possible approaches to reducing the imbalance but made no specific recommendations about which to pursue. Rather, it was an attempt to create the sort of shared understanding of the resources of the Colorado River Basin so essential to the rule changes that everyone knew would be needed. It was to provide, in Connor's words, "a common technical foundation . . . upon which continued dialogue will be built towards actions that will enhance and preserve the future of communities, economies, and ecosystems supported by the Colorado River."[2]

Like the Environmental Impact Statement that accompanied the 2007 Interim Guidelines, the Basin Study considered a range of hydrologic assumptions, demonstrating the increase in sophistication in our understanding of the Colorado River's hydrology. It included the long-term natural flow record (1906–2007), two tree-ring-based paleohydrologic methods, and estimates of system flows generated by climate change models. The study advanced science by investigating the future impacts of climate change on irrigation requirements of crops and native vegetation. It analyzed new and innovative agricultural and municipal conservation strategies. Further, breaking with the exclusive club approach that dominated past planning studies, the basin study process was transparent, and its committee structure included a broad range of basin stakeholders, including not just the states and major water agencies but also representatives from tribes, environmental groups, recreation interests, and smaller municipal and irrigation entities.

But that inclusiveness came at a price that undercut some of the report's technical honesty. Where ambiguities or long-standing differences in the governing rules had major hydrologic implications, such as the Upper Basin's 1922 compact obligation to Mexico, Upper Division state participants demanded the study fudge the math. For example, even though the minimum objective release from Lake Powell has been 8.23 million acre-feet per year for nearly five decades, the study only showed the hydrologic risk of the Upper Basin not making a 7.5-million-acre-foot release. The study only recognized the 75-million-acre-foot requirement under Article III(d) and ignored the disputed obligation to Mexico under Article III(c)—and the reality of how the river is operated. Further, when model results showed the actual flows at Lee Ferry dropping below 7.5 million acre-feet per year, the states forced the Bureau of Reclamation's modelers to inject what came to be called "magic water" into the model—extra water that existed in the computer model only, added to avoid compact shortfalls, thus inflating the water available below Lee Ferry (even when the cause of the low flow was climate change). In theory, this kept the study model results neutral on the unresolved legal issues. But the result, like so many times before in Colorado River Basin history, was a science distorted or ignored to meet political ends.

It also included a second flaw that would plague efforts in the years that followed to come up with the new rules needed to resolve the supply-demand imbalance. It failed to fully explore and independently document the major changes in demands that were already well underway, the economic disconnect between growth and water use that has been well documented in recent years. The data that were used to develop the different demand scenarios were provided by the states. In almost every case and in a virtual repeat of what happened during the first week of the negotiations of the 1922 compact, the individual states provided demand estimates that were based more on aspirations and the concept of "protecting their entitlements" than the actual reality of what was happening back home.

The resulting graph, showing a gap between supply and demand, quickly became iconic. Even after it became clear that demand was not rising the way the graph had projected, the image it left in the public mind of a region running out of water persisted.

Drought Contingency Planning

Colby Pellegrino, an engineer/hydrologist and self-described "numbers geek," grew up watching her home town of Las Vegas grow into a booming metropolis of over two million people. As director of water resources for the Southern Nevada Water Authority, the intergovernmental authority that

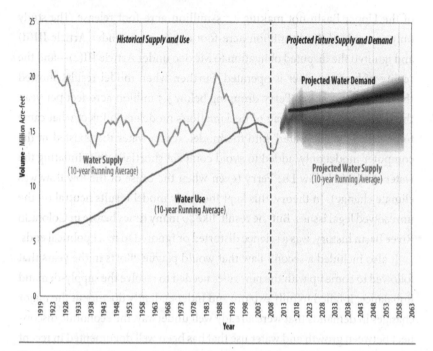

Figure 7 Historical and projected supply, use, and demand. Source: U.S. Bureau of Reclamation.

manages and distributes Nevada's 300,000-acre-foot apportionment of Colorado River water, Pellegrino now bears a major responsibility for ensuring that the taps still run for those two million people. In the summer of 2013, the Colorado River was experiencing another year of intense drought. Storage in Mead and Powell again plunged. That summer, Pellegrino prepared a number of model runs showing what would happen if the basin again experienced a 2000–2004 type drought. It was illustrative rather than predictive. Pellegrino was not saying we should expect a drought like that again, but rather raising a question. If a drought similar to the most severe in recent memory happened again, are we ready? The answer was no. Her results showed that without a large reductions in consumptive uses in both basins, another 2000–2004 drought would drain both Lakes Mead and Powell down to "dead pool"—the level at which water can't reach a dam's outlet works.[3]

Similar modeling results, showing that the modern Colorado River system would go bankrupt under the reoccurrence of past droughts had been presented at conferences or in basin-wide studies,[4] but the Pellegrino results were the first to trigger focused action by basin. In June 2013, representatives of the seven basin states met with Interior secretary Sally Jewell in Park City, Utah. At that meeting the secretary and the state representatives all agreed that the Colorado River Basin was not prepared for a continuation of the drought hydrology that had begun in 2000. They agreed that it was time for each basin to prepare drought contingency plans.

The work proceeded with a backdrop of science that was making it increasingly clear that the impact of climate change on the Colorado River was no longer some abstract future projection. In March 2016, the University of Arizona's Connie Woodhouse and her colleagues published a paper showing that warming temperatures were already reducing the Colorado River's flows. The work was followed soon after by a second paper by Colorado State University's Brad Udall. The work made it clear that what water managers thought they had been seeing—less water in the river for a given amount of winter snow in the mountains—was now a reality.[5]

While the scientific consensus about the impact of climate change on the Colorado River was clear, the scientific modeling needed to translate its implications into the kind of usable streamflow forecasts needed to model the future in a way that would support near-term policy decisions was not yet mature enough for practical use. Recognizing that fact, the river's managers began turning the old problem of "period of record" on its head. In the past, the early development community had repeatedly overestimated the amount of water available for Colorado River development by choosing an unusually wet period of record for their analyses of the Colorado River's flows. Now, they began doing the opposite. Much as Pellegrino had zeroed in on the possibility of a repeat of the 2000–2004 drought, the new approach focused on the period from 1988 to the present, reasoning that it more accurately represented the river's flows in a climate change world.[6]

The acknowledgment by the states that they needed to prepare plans that, when put in place, would reduce consumptive uses throughout the basin was, in some ways, code speak for acknowledging that the basin was now overdeveloped.

For the Upper Basin, that included a growing recognition that future development—the dreams of developing a bigger piece of the compact's

mythical 7.5 million acre-feet per year—was no longer possible, that growing development would inevitably come at the expense of increased risk to
current water users.

The process was the latest in a series of actions beginning in the 1960s that
were needed to come to terms with the reality that, in failing to heed LaRue,
Stabler, and Sibert, the basin's developers had failed to write rules governing
how to cut use if water supplies fell short.

The first step came in 1968, with the CAP authorization act giving the
project junior status if supplies fell short. Then, the 2001 Surplus Guidelines
recognized that the basin needed to wean California from its dependence on
surplus. The 2007 Interim Guidelines created a framework for implementing
Arizona's junior status. But none of these steps had gone far enough. Water-
using communities still hung to a dream of full canals and future development. Pellegrino's analysis showed that, after nearly a century of development based on optimistic hydrologic assumptions and illusions of importing
water from far-away rivers, the basin now had no choice but to move from
the era of development to one based on sustainability.

Because their situations were very different, the Upper Basin and Lower
Basin each pursued their own drought contingency plan on separate tracks.
For the Upper Basin, already living within its Colorado River Compact share,
the steps required were more modest—expanding the way the upstream reservoirs are operated to maintain the levels of Lake Powell and the possible
creation of a "demand management" program, coded language for paying
water users, primarily farmers, to use less. The decentralized nature of Upper
Basin water management and use makes problem-solving different from in
the Lower Basin. Water begins in snowpack and flows in small creeks, where
it is used along the way by numerous individual users and small ditches
interspersed with a handful of larger projects built by the Bureau of Reclamation and the region's larger municipalities. The Upper Basin has no centralized authority in charge of turning down the tap. Each state has its own
state engineer and differing rules, making water use reductions a challenge.

The centralized nature of Lower Basin management made the task more
straightforward, with the water released en masse from Hoover Dam to a
small number of major users under contracts with the secretary of the interior. But while the governance structure is simpler, the size of the changes

needed was challenging, requiring rewriting rules about how much each state's share would have to be reduced below the allocations made under the compact, the 1928 Boulder Canyon Project Act, the Supreme Court's decision in *Arizona v. California*, and the 1968 Central Arizona Project authorization. California expressed a willingness to give up the guarantees its senior status under the 1968 authorization had given it, while Arizona water users struggled to come to terms with the reality that a full Central Arizona Project canal was no longer a year-in, year-out certainty.

All of that was needed to put into place the next steps to unwind the mistakes embedded in the Law of the River by the failure to acknowledge the findings of LaRue, Stabler, and Sibert.

CHAPTER 20

Finally Listening to LaRue?

"The question," Bureau of Reclamation Lower Colorado River Regional Director Terry Fulp asked a packed Arizona auditorium in the summer of 2018, "is—what is the future going to look like?" It is the question that has bedeviled Colorado River managers since Eugene Clyde LaRue first measured the river's flow in the early years of the twentieth century. But in a striking break from the long-standing tradition of rosy scenarios to underpin critical Colorado River policy decisions, Fulp offered a sobering reality.

No longer could the river's management community promise, as LaRue's nemesis Arthur Powell Davis had during the 1922 compact negotiations, an 18.1-million-acre-foot-per-year river at Lee Ferry. No longer could the river's management community feint, as Floyd Dominy had in 1968 as it became ever clearer that the Colorado was a smaller river, with the promise of "augmentation" to bolster its shortcomings.

No longer could the Bureau of Reclamation and state water managers, as a team led by Fulp himself had done a dozen years earlier, lean on model results based on the premise that the last one hundred or so years of the river's flow was a reasonable representation of the future. It no longer made sense to start the analysis with the wet years of the first two decades of the twenty-first century, ignoring the droughts of the nineteenth century.

Things were bad enough when Reclamation used the full record available. From 1906 to 2016, Reclamation's Natural Flow Database showed the Colorado River was just a 14.8-million-acre-foot-per-year river at Lee Ferry, not enough to provide all the water allocated by the 1922 Colorado River Compact and the 1944 treaty with Mexico. With the Upper Basin using some 3 million acre-feet less than its 7.5-million-acre-foot compact apportionment, the basin might have been able to cope with that mistake. But by the summer of 2018, the river's managers were forced to come to terms with the fact that even the relatively pessimistic 1906–2016 record commonly used to calculate the risks of shortage was too rosy. "All indications are," Fulp said, "that the past 109 years is not the best representation of the future." The historic record might suggest a 14.8-million-acre-foot-per-year river at Lee Ferry, but the Colorado was no longer even that.

Fulp was encouraging his Arizona audience to finalize the state's internal plan to reduce its use of Colorado River water. But he knew that Arizona's plan, part of a broader effort among the seven basin states to write new rules, was only one step up a long, steep hill. The 2007 Interim Guidelines expire at the end of water year 2026 and the negotiations necessary to put in place the rules on how the river would be operated after 2026 were scheduled to begin no later than 2020.

Negotiating the post-2026 river will be one of the most challenging tasks facing basin water agencies, state and tribal water officials, recreation and environmental interests, and representatives of the federal government. They have the potential to rival the 1922 compact negotiations in terms of both importance and drama. Some of the fundamental issues facing the negotiators in 2020 are similar to those Hoover, Carpenter, Davis, and their contemporaries faced a century ago:

How much water will the Colorado River provide in the future?
What are the legitimate demands?
How should the basin be governed under conditions of uncertainty?

But underlying the three basic questions are several important realities. First, every drop of Colorado River water is now fully used and spoken for. There is no longer a split between the faster- and slower-growing states.

Instead, it is between the haves and have-nots. The haves are primarily senior agricultural users and the have-nots are urban areas with primarily junior rights but a need for certainty of supply, along with Native American communities and the environment. Second, climate change is drying out the Colorado River Basin and those areas surrounding it that, through exports, use and rely on Colorado River water. Finally, the basin is much more complex and diverse. Its cities are booming. Most all of the basin's residents value the wonders, beauty, and diversity of the Colorado River Basin and demand that their governments take care of it. We now recognize that Native American communities deserve a rightful share.

The Future Water Supply

Perhaps the most uncertain question facing the basin is the future of the river's flow. The technical details of what Fulp and his colleagues did are important. Beginning in 2013, a group of water managers in the Colorado River's Upper Basin had begun advocating for the development of a new, more realistic assessment of the Colorado's hydrologic risks. With a warming climate, they argued, using the statistics of the past to represent a plausible future no longer made sense. They could already see a warming climate sapping the river's flow and were looking for an analytical tool that could represent the changing reality. Lacking confidence in the ability of science's best climate change models to accurately represent the fine-scale details of the river's flow, they cast about for something else. Rather than focus on the entire 1906–2015 record, they argued, planners should focus on the last few decades, as the climate warmed and the river's flow dropped. They called it "the stress test." It uses the same modeling tools with which the river basin's managers were already familiar. But rather than using the statistics of the entire record to model plausible futures, they zeroed in on the period 1988 to the present. During that time, the river's flow was over 10 percent less than the 1906–present average. The Colorado should no longer be considered the river the authors of the Law of the River thought could deliver 17.5 million acre-feet per year with a surplus left for future allocation or even the 15-million-acre-foot river that we envisioned in writing the 2007 Interim

Guidelines. Moving forward we should plan on no more than a 13 million acre-foot per year river.[1] The stress test was the vehicle to do this.

As Fulp spoke, the combined storage of the reservoirs behind Hoover and Glen Canyon Dams sat at 21.2 million acre-feet, the lowest since Lake Powell was first being filled in 1966. That was down 26 million acre-feet since they were last full two decades before. No longer was it possible to, as the late Colorado River District general manager Rolly Fischer had joked, promise the communities more water than was available in the river, then hope that a future generation would fix the mess. Using the new, more realistic "stress test" hydrology, Fulp explained, there was a nearly fifty-fifty chance that by the mid-2020s, even with the water delivery cutbacks already envisioned by the shortage sharing agreement signed by the seven basin states in 2007, Lake Mead's surface elevation could drop below 1,025 feet, a level at which there was less than a year's water left and no clear plan to manage the deep cutbacks that would be required. Fulp's message was clear. The basin desperately needed plans to manage what followed. That required immediate steps to reduce water use now, a drought contingency plan. Such a plan required a more honest assessment of how much water the river actually had in order for water managers within the states to know how much they would need to cut back.

A major challenge facing the negotiators of the post-2026 river rules will be coming up with a realistic assessment of the river's future flow. For this task, even the 13.1-million-acre-foot natural flow average from the stress test may be too optimistic.[2] Too many of the basin's state and federal water agencies have been unwilling to acknowledge that the river flow conditions experienced since the late 1980s are likely the new normal and not an "extended drought," as they refer to in their talking points. We can only recommend the designated negotiators follow the advice of LaRue and Hundley by reaching out and listening to the experts, then using Sibert's precautionary principle to look at the river under a realistic assessment of future flows.

The traditional story told about the allocation of the Colorado River's water is that the compact negotiators did the best that they could with the limited hydrologic information available in the 1920s. This makes those early decision-makers seem simple victims of a short period of record that was unfortunately very wet. But that lets them off the hook too easily. In choosing Arthur Powell Davis's numbers when they overallocated the Colorado River's water, they willfully ignored the available data on a longer and therefore more robust period of record. LaRue, Stabler, and Sibert all had offered a more realistic assessment, with strong science and a good rationale for taking a more conservative approach.

What can science tell us today about the hydrologic reality of the Colorado River Basin, and how can we use it?

The first point is one E. C. LaRue made more than a century ago—that the river's flows are subject to huge variability. LaRue's paleoclimate reconstruction, using the levels of the Great Salt Lake to infer past droughts, may have been inconvenient or unpersuasive. But a century of operating experience, and the use of sophisticated tree ring chronologies to peer into the past, leave no doubt that deep decadal-scale droughts, and persistent wet periods, are a central feature of the Colorado River Basin's hydrology. Those swings are why Glen Canyon Dam was built in the first place—a lesson from the dry 1930s about the importance of full storage going into a drought.

The second point, also suggested by LaRue a century ago, is that independent of variability, there is simply less water in the system, on a long-term average basis, than the compact's framers thought. Arthur Powell Davis's 18.1-million-acre-feet-per-year natural flow at Lee Ferry has been, in reality, 14.8 million acre-feet—18 percent less water, year in and year out, than the early developers planned for.

The third point is that rising greenhouse gases are inexorably changing the climate. The projections of reduced Colorado River flow as a result of climate change that began in the 1980s[3] are no longer projections. The observations of water managers that they are seeing less water in the river relative to the snow falling in the Colorado's headwaters, as warmer temperatures lengthen growing seasons and increase evaporation and plant water use, have been borne out by scientists. Since the late 1980s the river's flow has dropped to just 13.1 million acre-feet per year, 28 percent less than the Arthur Powell Davis's 18.1 million acre-feet, the unrealistic premise that underlies the allocation of the river's water.[4]

The future has arrived, and it is a stark one for the river's managers and the communities that depend on them.

Decoupling—Future Demands for the Colorado River Water

There also is an empirical reality on the water demand side that has been insufficiently taken into account, but which provides a significant opportunity to help solve the river basin's problems once we are willing to take it seriously. The widespread presumption that population growth means growing water demand drives much of the politics of water planning in the Colorado River Basin. But it is wrong. Simply put, we are consistently using less water. In almost all the municipal areas served with Colorado River water, water use is going down, not up, despite population growth. Water use in the basin's major agricultural regions also is going down, even as agricultural productivity continues to rise. This is not limited to the Colorado River Basin. Such "decoupling" between water use, population, and economies is common across the United States.[5]

We have been getting this wrong for a century. When representatives of the basin's states opened the Compact Commission discussions in 1922 with a claim that they expected 10 million acres to eventually be brought under irrigation, they were vastly overestimating the basin's agricultural potential. Today, the actual number is less than five million acres.

A similar error continues to plague attempts to solve the basin's water problems. Overestimating future water need and underestimating communities' changing behavior in response to scarcity is deeply embedded in the culture of Colorado River water management. The Bureau of Reclamation's 2012 basin study projected a 3.2-million-acre-foot imbalance as demand rose inexorably. But even as the projections set into motion a scurry of long-range planning exercises to deal with the problem, actual water use was headed in the opposite direction. In the first decades of the twenty-first century, Imperial Valley irrigation declined by more than fifty thousand acres, and total annual water use there dropped half a million acre-feet. Municipal water use in the state's sprawling coastal plain was headed in the same direction, with total demand at the levels it had been in the 1980s, even as the region's population had grown 57 percent.[6] While the 2012 basin study had projected that California's demand would rise to 5 million acre-feet per year by 2015, in fact it was headed in the opposite direction, dropping to 4 million acre-feet by 2017.[7]

The same can be seen in the Upper Basin, where in the early 1980s Denver Water was serving about 250,000 acre-feet of water per year to 800,000 people. Today it is serving close to 1.5 million people with about that same 250,000 acre-feet.

This is a pattern repeated again and again, as municipal water use declines in nearly all of the major metropolitan areas served by the Colorado River, from the Los Angeles-San Diego region, to the glittering fountains of Las Vegas in Nevada, to Phoenix and Tucson in Arizona, to Albuquerque and Santa Fe in New Mexico.[8]

Despite this success, demand still outpaces supply. You can see this in the basin's declining reservoirs. But the basin's major municipalities have shown that growth and economic prosperity can occur without necessarily increasing total water consumption. The basin's municipalities need to acknowledge that, with a few exceptions, they can live, and live well, with the water they currently have and not seek new supplies. Instead their focus and priority will be to cooperatively work with senior users to provide additional certainty to their existing supplies.

Even as basin managers need to learn from the mistakes of the past about incorporating good hydrologic science in their planning, they also need to heed mistakes in projecting how much water is needed in the future.

The Future of the 1922 Compact and Basin Governance

The 1922 Colorado River Compact was far from perfect. There were still many controversies, and litigation was still a necessary and important tool to settle differences among the states. Science was too often ignored. The amount of water in the river was routinely overstated—too often because it suited the politics of the moment. But despite its flaws, the 1922 compact and the laws, treaties and other compacts that followed it worked because they were based on a shared understanding of the mutual needs, mutual risks, and mutual benefits of different communities across the basin whose needs might otherwise have been in conflict. Today, the river's waters are fully developed and consumed. But this has come with a cost. Rarely since the late 1990s has a drop made it to the Gulf of California.

Today, the challenges are sustainability and certainty of supply in the face of the deep uncertainty created by climate change. The impacts of climate change will increase uncertainty by elevating demands, reducing stream flows, and altering the timing of the river's flow. Meanwhile economic and population growth means the pressure to continue the region's water conservation success to bring supply and demand into balance will grow. Maintaining water quality and aquatic ecosystem health will be more challenging, and recreation pressures and uses will continue to grow. Because irrigation users were the first to develop, they have the senior rights. The region's now booming urban areas have the junior and thus most insecure water supplies, the ones that are subject to the first and largest shortages if the water is not available. The conflicts between those who have the most secure rights and those that must have certainty to meet the needs of people and the water supply uncertainties due to climate change are the underlying threats to stability.

In the next hundred years, if the 1922 compact is to continue to be the bond that holds the Colorado River family together, to resolve controversies and minimize litigation as intended by its founders, the most important social contract will no longer be between the faster- and slower-developing basins. While that Upper Basin-Lower Basin social contract will endure, embedded as it is in both the customs and the rules of the river's management, the new social contract must be among the senior agricultural users and the more junior cities that need certainty of supply. Instead of a blueprint for development, the basin is now in need of a stable and effective governance of the use of the river's waters under conditions where current demands already exceed the existing supplies and where future supplies will be both variable and uncertain and with mechanisms in place to move water among uses.

Like one hundred years ago, the river's future is not all dark. Innovation, cooperation, and an expanded reliance on science are now the foundation for basin-wide solutions. The allure of interstate litigation is slowly fading from the basin's future (except perhaps in the basin's statehouses and law schools). Water agencies fully understand that litigation will not create one

drop of additional water and instead will undermine the kinds of cooperative solutions needed to solve the basin's water problems. Economically,
there has been a decoupling between the consumption of water and regional
economic output. Finally, the water management family has become more
inclusive. Until recent decades, all the major development decisions were
made by a small number of individuals and agencies. Although many of
the decisions received considerable public attention and press coverage and
often high-level officials such as governors and senators were personally
involved, input from those outside the small water community was rarely
sought or accepted. Today, the process is much more open and inclusive,
and the basin has benefited from a broader range of ideas, especially better
and more diverse science.

Former Southern Nevada Water Authority general manager Patricia
Mulroy is often quoted as saying the Colorado River Compact can mean
anything the parties to the compact agree that it means.[9] This truism has
always accompanied the compact. When Arizona refused to ratify the compact, the remaining states adopted a six-state ratification strategy and the
compact survived. After the Sibert board report raised serious questions
concerning its hydrologic underpinnings, its founders pushed on. When
Congress refused to subordinate navigation to other uses as contemplated by
the compact, it survived. The compact survived legal challenges in the 1930s
and again in the 1950s when the Supreme Court shielded it from challenge
by limiting its decision in *Arizona v. California* to interpreting the 1928 act.
In 2005, when the Upper Basin states challenged Secretary Norton to reduce
Glen Canyon Dam releases from 8.23 million acre-feet per year to 7.48 million acre-feet per year, cooler heads (aided by a wet year) prevailed, and the
basin states backed away from litigation.

In many ways, the basin states are in a similar position to that of the late
1920s. In 1927, six states chose to continue the ratification process of the
imperfect compact they had negotiated rather than reopening the negotiations to appease Arizona and face an unknown future. One of the more
likely outcomes was a future without a compact, an unacceptable situation
to all of the states but Arizona. Today, the choices are to continue with the
1922 compact as written, with its embedded flaws but open to broad interpretation by the parties, or to engage in major litigation at the U.S. Supreme

Court. The outcomes of litigation could include a future without a compact, a future where the compact has been so narrowly interpreted limiting future cooperative and innovative solutions, or even a future where the river is "federalized" far more than it is today.[10] All of those are unacceptable to most, if not all, of the basin states. This means that while litigation is always possible, the inherent risks and ensuing chaos are so great that its chances are much diminished.

It is within that negotiating framework that the work ahead must be done. Moving forward with a post-2026 river under the umbrella of the 1922 compact will require a recognition and at least a tacit acceptance of those provisions that are inconsistent with the river as it is managed today. There are also several provisions where there has never been a resolution or even common understanding within the basin states. Unless these conflicted provisions are carefully addressed and creative ways found to resolve or work around them, the basin is in for difficult times.

The first major issue is the development inequities within the basins. Existing and foreseeable consumptive uses in the basins do not match the Article III(a) and (b) compact apportionments of 8.5 million acre-feet to the Lower Basin and 7.5 million acre-feet to the Upper Basin and likely never will. Counting Lower Basin tributary use, evaporation, and system losses, total annual consumptive uses in the Lower Basin are in the range of 10.5–11.5 million acre-feet per year, 2–3 million acre-feet per year above its compact apportionments.[11] Uses in the Lower Basin peaked in 2002, the year before California was forced to reduce its uses to 4.4 million acre-feet per year. Out of necessity, since 2002 consumptive uses in the Lower Basin have continued in a downward trend. In the Upper Basin, current consumptive uses, including reservoir evaporation, are in the range of 4.0–4.5 million acre-feet per year, 3.0–3.5 million acre-feet less than its apportionment. Further, since the late 1980s, growth in consumptive use in the Upper Basin has leveled off.[12]

Moving forward with this reality means that the major burden of reducing uses will be in the Lower Basin. The Lower Basin will have to continue its efforts to reduce the structural deficit. Without a doubt, there are still development expectations and pressures in the Upper Basin; while the Lower Basin is reducing uses, the Upper Basin may be increasing uses. However, the net increase from new or expanded projects will likely be small; the major increase in consumption may be the effect of rising temperatures on existing irrigated crops.[13]

The second major and related issue is that because the Upper Basin states have fixed obligations under Articles III(c) and (d) at Lee Ferry, these states bear a much greater climate change risk.[14] Despite the reality that it is critical to the determination of how much water must be released from Lake Powell to the Lower Basin, the policy differences over the interpretation of the Upper Basin's obligation to Mexico, Article III(c), have been ignored or finessed for decades.

The Grand Bargain

The combination of the Lower Basin's current consumptive use of more than 8.5 million acre-feet, the disputed issues related to the Upper Basin's obligation to Mexico, and the need for both basins to share the risks of future climate change may open the door for what has been called the "Grand Bargain." To get to a Grand Bargain will require both creative leadership from all stakeholders (not just the states and Bureau of Reclamation) and a reliance and trust in science that has so far eluded the basin's water decision-makers. That trust requires a recognition and acceptance of the fact that there is not enough water to do all the things people want.

In our view, the Grand Bargain could be a conceptual agreement under the umbrella of the 1922 compact.

The basin stakeholders would recognize that enforcing or litigating the letter of the law (the 1922 compact) would result in havoc in both basins. Municipalities like Denver in the Upper Basin cannot live with the uncertainty created by the threat of a Lee Ferry curtailment. Municipalities in the Lower Basin, including the Arizona cities served by the Central Arizona Project, the Southern California cities served by the Metropolitan Water District, and Las Vegas, cannot live with the consequences of a lower-basin limit of 8.5 million acre-feet, including reservoir evaporation and tributary use.

Without a basin-wide approach, cities will be forced to be turn to their agricultural neighbors, aggravating the urban-rural conflict.

While everyone has water lawyers ready to argue that the havoc should not settle upon them, there is not enough water in the Colorado River for all the lawyers to be right.

The alternative—the Grand Bargain—must be based on a recognition that the goal of river management and conservation programs must begin with the protection of all existing uses in the basin, including recreation and environmental flows.

The key to a compromise is for the Lower Basin to remove the threat of a curtailment of Upper Basin water use to meet Lee Ferry delivery obligations in return for the Upper Basin's recognition of the existing level of development and use in the lower basin. This would spread the risk of climate change across all the basin's users.

The process by which such a grand bargain might happen may be every bit as important as the technical details of what it would entail. At a 2005 meeting of the "basin states principles"—the Colorado River leadership team representing each of the seven basin states—representatives from Colorado floated a proposal. The details involved some tricky trade-offs. But the details are less important than the forum.[15]

Such an agreement cannot be specified ahead of time but has to emerge from the process of collaboration and compromise that has grown up over the last two decades. That 2005 meeting is an example of the sort of meetings that happen all the time, as representatives of the basin water community meet to hash out their problems.

The conflicts between regions and groups of water users are real, but the old myth that "water is for fighting over" has given way to a process that has laid the foundations for the kind of agreement that will be needed to reach a grand bargain. It is through this sort of long-term process that a sustainable system of Colorado River governance must emerge.[16]

Epilogue

"An Unknown Distance Yet to Run"

In his account of the first expeditions of the Colorado River by the continent's European immigrants, John Wesley Powell sat at the confluence of the Colorado and the Little Colorado Rivers, in the deepest reaches of the Grand Canyon, and pondered the uncertainties that lay ahead:

> We are three quarters of a mile in the depths of the earth, and the great river shrinks into insignificance as it dashes its angry waves against the walls and cliffs that rise to the world above; the waves are but puny ripples, and we but pigmies, running up and down the sands or lost among the boulders.
>
> We have an unknown distance yet to run, an unknown river to explore. What falls there are, we know not; what rocks beset the channel, we know not; what walls rise over the river, we know not. Ah well! We may conjecture many things.[1]

It is a nineteenth-century literary flourish often quoted today, because much of what Powell said rings familiar to the river management community of the twenty-first century—the unknown distance yet to run, with no way to know where the falls are, with the certainty that there will be unexpected rocks in the channel. But we have learned a great deal about the Colorado River and our relationship with it as humans now so dependent on its water.

When E. C. LaRue wrote in 1916, before the Colorado River Compact, before Hoover Dam, before all the development that was to follow, that "the flow of the Colorado River and its tributaries is not sufficient to irrigate all the irrigable lands lying within the basin," he prefaced his observation with a caveat. "More complete data," he wrote, "would probably indicate a *greater* shortage in the water supply available" (emphasis added).[2]

Each of those points has largely been lost to the history on which our modern operational understanding of the Colorado River Basin is based. The first, as we have seen in the stories of LaRue, Stabler, Sibert, Tipton, G. E. P. Smith, Stockton and Jacoby, and the modern climate change scientists is surely the most important. There is simply less water in the system than the edifice of laws and policies and infrastructure was premised upon. But LaRue's second point may be the more important—the need for humility in the face of uncertainty, and the crucial need to design that humility and uncertainty into the institutions we build to use and manage the river.

"To study the past," the historian Jill Lepore has written, "is to unlock the prison of the present."[3] The conventional story of the great mistake of the overallocation of the waters of the Colorado River—that the authors of the Law of the River did the best they could with the limited data they had, unaware that they were living in unusually wet times—is the foundation of such a prison. The "How could they have known better?" story, in absolving Herbert Hoover, Delph Carpenter, Arthur Powell Davis, and the others, obscures essential lessons.

The first lesson is the most obvious. When we sent LaRue Stabler, and their colleagues into the field in the first quarter of the twentieth century to gather the best available information about the flows of the Colorado River, we should have paid heed when they returned with the answer that the river's flow was "not sufficient" for the works on which we were about to embark.

There was a tendency, which we sometimes share today, to seek out a single number to represent the "flow of the Colorado River" to use as the benchmark against which to manage the water's use. To the extent we do that, the first lesson of LaRue, Stabler, and the others is to use the best data, not the data which best supports one's hopes and dreams.

But even more than being honest about the data, LaRue's caveat in 1916 about the incompleteness of the data goes the farthest to helping explain the prison in which we now find ourselves. *Humility* is a word likely not used often in descriptions of Eugene Clyde LaRue. The arrogance of a man willing to brashly interrupt U.S. senators as he tried to persuade them of the idiocy of the plan on which they were embarking offers an easy explanation for why his ideas were shunted aside. But to read the care in his work, as he explained the deep uncertainties in the numbers he was offering, is to see a genuine humility in the face of the difficult task of determining how much water had flowed in the Colorado River in the past and what that might tell us about how much we might expect in the future.

There are missed opportunities aplenty in the narrative of the Colorado River's management over the course of the twentieth century, "if only they had listened" moments when yet another expert pointed to the building problems. But perhaps the most important opportunity lost came in 1928, when members of Congress in 1928 chartered a commission to report back on the feasibility of the Boulder Canyon Project Act.

This was the moment of commitment, when the plans for the development of the Colorado River were moving from abstract discussions in commission meeting rooms and Senate hearing chambers to poured concrete and diverted water. This was the moment Congress turned to independent experts to give the enterprise one last look before forging ahead. For that reason, the words of William Sibert, the army general brought back from retirement to offer his expert advice, are worth careful notice.

Painstakingly, Sibert explained the uncertainties—the shortcomings of the Yuma gauge, the difficulties in estimating flow based on older measurements taken at the Yuma railroad bridge, the reconstructions based on measurements taken on the river's upper tributaries, the inferences to be drawn from the rise and fall of the levels of the Great Salt Lake, the actual measurement of rainfall to the north, from whence the Colorado River's water flows.

Sibert's bottom line reflects a profound humility in the face of the limitations of the science and the data of the day. "The records of past performance of the Colorado River and of such other streams in this vicinity as seem pertinent furnish no basis for an exact estimate of long-past flows in the Colorado River." And therein lay the risk. "There is naturally considerable leeway in the interpretation of these data," Sibert wrote, "and estimates based thereon may differ materially."[4]

The difference in available estimates offered the opening Arthur Powell Davis needed for the optimistic numbers in the Fall-Davis report and his enthusiastic assurances going into the Colorado River Compact negotiations and the ensuing years needed to win approval of the ambitious projects to follow. "They had an opportunity to pick and choose information that best suited their interests and uncertainties," as Norris Hundley wrote. "And that is what they did."[5]

Sibert, trained in the classically conservative methods of an engineer, urged a different approach to the uncertainties, arguing for using "estimates . . . on the safe side." Just as engineers build safety margins into their structures, Sibert was arguing for a caution in the development of the Colorado River's waters.[6]

But beyond the specific numbers to be used in estimating the river's flow and the resulting development plans, the numbers of LaRue, Stabler, Sibert, and the others suggest the need for a second sort of humility that may be the most important lesson we can learn from this history. It is the need to imbue the institutions for allocating and using the Colorado's waters with the flexibility to respond when there is less water than we planned.

Much of the increasingly elaborate institutional architecture of river management in the twentieth century—the arcane debates over "salvage by use," the unresolved questions about accounting for reservoir evaporation and Mexican treaty delivery obligations, the legal uncertainties about when and how Lower Basin water users might be cut off as Lake Mead drops—was an effort to maintain a fiction of plenty of water in the face of a reality of less. The most profound failure of the river management institutions of the twentieth century lay in the lack of a plan B. What if there is less water?

That is the challenge going forward. There is, in fact, less water. Our best modern understanding suggests LaRue, Stabler, and Sibert were right—that there probably always was less. But with climate change, there clearly is less now, and the best science of the first decades of the twenty-first century suggests we don't know how far below us the floor lies.

ACKNOWLEDGMENTS

The authors would like to thank the many people in the Colorado River and western water communities who have helped and supported our efforts to understand these issues: John Carron, Jim Pokrandt, Andy Mueller, Peter Fleming, Dave Merritt, Dave Kanzer, Laurie DePaolo, Sara Porterfield, Hannah Holm, Brad Udall, Jack Schmidt, Mike Cohen, Kathryn Sorensen, Bill Hasencamp, Colby Pellegrino, Bart Fisher, Tanya Trujillo, Anne Castle, Doug Kenney, Tom McCann, Chuck Cullom, Terry Fulp, Bob Snow, Jim Prairie, Carly Jerla, Jim Lochhead, Pat Mulroy, and Amy Haas.

Eric's journey with this book began on a river trip when two talented journalists, Heather Hansman and Sarah Tory, suggested he turn his 2007 report, *The Colorado River: A Quest for Certainty on a Diminishing River*, into a book.

Members of the University of New Mexico water community, especially Bruce Thomson, Bob Berrens, Becky Bixby, and Julie Coonrod, have provided a welcoming home for John's continuing efforts to understand and explain the waters of the West.

Kerry Smith's editorial grace and the insights of two anonymous reviewers vastly improved the manuscript.

Eric would like to thank his family, wife Sue—hoping someday her husband might read a book about something other than the Colorado

River—daughters Kenzie and Hallie, and son-in-law Carl, and a special acknowledgment to his father Richard, a graduate of Needles High School, class of 1941, who remembers the Colorado River before dams.

John would like to thank his family, Elizabeth and Bob Fleck for first taking him to the rim of the Grand Canyon to look down on the Colorado River, Reed and Lisa for their patience when he seemed so distracted, and Lissa for cheerfully embracing, supporting, and joining in the journeys.

NOTES

Introduction

1. Hundley, *Water and the West*, 352.
2. Hundley, *Water and the West*, 352.
3. Reisner, *Cadillac Desert*, 271.

Chapter 1

1. Brean, "Drought."
2. LaRue, *Water Power* (hereafter cited as WS556).
3. Clarke and King, *Water Atlas*.
4. "Canyon Flood Peril Pictured"; Boyer and Webb, *Damming Grand Canyon*, 242.
5. WS556, 123.
6. There are 1,233.5 cubic meters in an acre-foot of water.

Chapter 2

1. Ives, *Report upon the Colorado River*.
2. Midvale, "Prehistoric Irrigation."
3. Summit, *Contested Waters*.
4. Ives, *Report upon the Colorado River*, 73.
5. Wheeler, *Annual Report*.

6. Powell, *Report on the Lands of the Arid Region*, 8.

7. McGee, "Water as a Resource," 521.

8. Ross, *Promise of the Grand Canyon*, chap. 11.

9. Reclamation Act of 1902, Pub. L. No. 57–161, 32 Stat. 388 (1902).

10. Before settlement, the Imperial Valley was referred to as the Colorado River Desert. It was changed to the Imperial Valley for marketing purposes.

11. U.S. Bureau of Reclamation, *Colorado River: Natural Menace*. Commonly referred to as House Document 419 or the "Blue Book" (hereafter cited as HD 419).

12. Once the river broke through the original control works that separated the river from the canal, it now had a gravity route via the man-made canal into the Alamo and New River channels, which drain northward into Imperial Sink.

13. HD 419, 59.

14. Tyler, "Delphus Emory Carpenter," 228. Tyler, *Silver Fox of the Rockies*.

15. *Kansas v. Colorado*, 206 U.S. 46 (1907).

16. *Wyoming v. Colorado*, 259 U.S. 419, 42 S. Ct. 552, 66 L. Ed. 999 (1922).

17. Mackey, *Protecting Wyoming's Share*, chap. 3.

18. Tyler, "Delphus Emory Carpenter," 237.

19. Tyler, "Delphus Emory Carpenter," 228.

20. Tyler, "Delphus Emory Carpenter," 241; HD 419, 60.

21. HD 419, 60.

22. The legal theory behind the concept of a compact is found in Article I of the U.S. Constitution.

23. Powell, *Report on the Lands of the Arid Region*.

24. Davis, *Irrigation near Phoenix*.

25. Newell, *Proceedings of First Conference*. Lippincott (1864–1942) had a fascinating and complicated engineering career. Kahrl, *Water and Power*, gives credit to Lippincott for identifying the Boulder, Black Canyon, and Parker Dam sites on pages 264–65.

26. LaRue, *Colorado River and Its Utilization*. Hereafter cited as WS395.

27. Until 1921, the Colorado River above the confluence with the Green was referred to as the "Grand."

28. WS395, plate XIX, page 192. Since there were no discharge records of the Colorado River at Yuma before 1902, to make his reconstructions back to 1895 he used records for the Green River, Grand River and, interestingly, the Arkansas River at Canyon City, Colorado, page 196.

29. The current stream gauge location, referred to as Lees Ferry, is just above the confluence with the Paria and at the historical location of the ferry. There is a separate gauge on the Paria River. To obtain the flow at the Lee Ferry compact point, one adds the Colorado's gauged flow at Lees Ferry to the gauged flow of the Paria.

30. In WS395, LaRue includes a table on page 194 showing the flow at points upstream of Laguna Dam as a percentage of the flow at Laguna. Using this table and adding estimated depletions upstream of Lee Ferry for 1895–2014, the estimated natural flow at Lee Ferry would be 16.2 million acre-feet per year.

31. LaRue understood the need to correct gauge flows for upstream depletions. In WS395, he raises the problem. In WS556, he corrects the flows at Lee Ferry to a common 1922 level of depletion.

32. See WS395, 190–214, for the discussion of river control (regulation). Boulder Dam is mentioned on page 180 in the power section where it states that "private parties contemplate constructing a dam to a height of 125 feet to obtain an effective head of 110 feet."

33. Mead et al., *Report of the All-American Canal Board*. The board's recommendations are found on pages 68 and 69.

34. Mead et al., *Report of the All-American Canal Board*, table 1, 17.

35. U.S. Reclamation Service, Problems of Imperial Valley and Vicinity, S. Doc. 142, 67th Cong., 2nd sess. (1922) (hereafter cited as "Fall-Davis report").

36. Simonds, *Boulder Canyon Project*.

37. LaRue was the primary champion of the Glen Canyon site, but he had supporters in Arizona.

38. Fall-Davis report, 18–20.

39. Fall-Davis report. Table 1 is on page 2, table 6 is on page 5. Another relevant table is table 9 on page 37, "*Averages 1903–1920*," of appendix B. This shows the water supply at the Boulder Canyon Project dam site.

40. Meko et al., "Medieval Drought."

41. Fall-Davis report, appendix B, "Water Supply and Development," 34–37.

Chapter 3

1. Davis, *Eighteenth Annual Report*, 407.

2. E. C. LaRue to Nathan Grover, April 7, 1920, Huntington Library LaRue Papers.

3. *Wyoming v. Colorado*.

4. The Laramie River has its headwaters in the mountains to the west and north of Fort Collins, Colorado. It flows out of Colorado on a northerly course into Wyoming to its confluence with the North Platte River near Wheatland, Wyoming.

5. Hoover was born in Iowa but considered California his adopted home. Hoover was not the first choice of everyone in the basin. Colorado's Carpenter wanted a high-profile international lawyer such as Elihu Root. Hundley, *Water and the West*, 130–31.

6. Arthur Powell Davis (1861–1933) was the nephew of John Wesley Powell. After graduating from George Washington University in 1888, he worked for the

USGS. He became chief engineer of the Reclamation Service in 1906, then director from 1914 to 1923.

7. Colorado River Commission, *Minutes and Record of the First Eighteen Sessions*, first meeting, 2.

8. Colorado River Commission, *Minutes and Record of the First Eighteen Sessions*, first meeting, 28.

9. Despite clear signals that cities, including Denver on the east and Los Angeles on the west, had plans in the works to export water from the basin, municipal use was not considered. Davis may have assumed that a division of water based on irrigated acreage was at least a nonarbitrary way of dividing the waters, and later, if the states wanted to use a portion of their supply for municipal uses, it was their choice.

10. While the Reclamation Service data was the most recent, it was not complete. An example of this is the state of Nevada. The minutes of the sixth meeting of the commission include five separate tables of data that were presented by the committee. The Reclamation Service data, table A, page 70, shows that Nevada was only using 2,000 acre-feet as of 1920 and had future uses of only 5,000 acre-feet. Davis acknowledged that they had incomplete information for Nevada.

11. The states also had their own data on existing uses, which on the whole were only slightly less than the Reclamation Service. Both the states and Reclamation used the same values for the amount of irrigated agriculture, but the states assumed a slightly smaller consumptive use per acre.

12. Colorado River Commission, *Minutes and Record of the First Eighteen Sessions*, sixth meeting, 74.

13. If one includes better data for the state of Nevada, then the state data was higher than the Reclamation Service data for the Lower Basin states by about 450,000 acre-feet. Still, this is only 10 percent greater than the Reclamation Service's estimates. In comparison, the upper states were collectively 100 percent higher than Reclamation.

14. Construction began on Laguna Dam in 1905 and was completed in 1910. The gauge on the Gila River at Dome was notoriously bad. From 1904 to 1914, the USGS did not have enough confidence in the measurements to report them in their annual water supply papers.

15. To complicate the problem, the 1902 gauge record is only nine months. The 1903–20 average is 17.4 million acre-feet per year.

16. At the eleventh meeting, Colorado's Delph Carpenter would propose a compact based on 17.4 million acre-feet at the Yuma gauge, believing he was splitting the river, but he did not correct the flow for the location of the IID diversion.

17. After the passage of the Reclamation Act in 1902, construction of irrigation projects accelerated, including the Salt River Project and Yuma Projects in

Arizona and the Uncompahgre, Grand Valley, and Strawberry Valley Projects in the Upper Basin.

18. The 1.3 million acre-feet is derived by correcting the flow by 800,000 acre-feet for upstream development on the main stem and 500,000 acre-feet for development on the Gila. The increase in consumptive use on the Gila between 1900 and 1920 is actually more like a million acre-feet per year, but the impact of a million acre-feet of upstream depletions on the gauge at Yuma would be about half a million acre-feet.

19. Fall-Davis report, table 8, page 36. This is actually close to the total Colorado River Storage Project initial unit reservoir evaporation in the Upper Basin and main stem reservoirs and system losses in the Lower Basin.

20. Colorado River Commission, *Minutes and Record of the First Eighteen Sessions*, sixth meeting, 80–81.

21. Hundley, *Water and the West*, 146–47.

22. Hundley, *Water and the West*, 149.

23. Nathan Grover was the head of the USGS Hydrologic Branch. He attended the initial meetings in Washington but did not attend the meetings in Santa Fe.

24. Colorado River Commission, *Minutes and Record of the First Eighteen Sessions*, sixth meeting, 70–71.

25. WS395, 168.

26. Colorado River Commission, *Minutes and Record of the First Eighteen Sessions*, seventh meeting, 124.

27. Carpenter believed exports or transmountain diversions from the Upper Basin would not exceed 500,000–600,000 acre-feet per year. See Tyler, *Silver Fox of the Rockies*, 169 and 174. Today total exports in the Upper Basin are in the range of 800,000 to 1 million acre-feet per year.

28. Hundley, *Water and the West*, 138.

29. Colorado River Commission, *Minutes and Record of the First Eighteen Sessions*, seventh meeting, 151.

30. Colorado River Commission, *Minutes and Record of the First Eighteen Sessions*, seventh meeting, 148–52.

31. Throughout the minutes, there are suggestions to make the commission permanent and use it to address problems and avoid future controversies like the sharing of shortages. Carpenter was opposed to this idea. Two decades later the states with Upper Basin interests would adopt this model for the Upper Colorado River Basin Compact, which provides for a permanent commission.

32. Hundley, *Water and the West*, 167–79.

33. Hundley, *Water and the West*, 180.

34. Hundley, *Water and the West*, 182–83.

35. Colorado River Commission, *Minutes and Record of the First Eighteen Sessions*, eleventh meeting, 23; Hundley, *Water and the West*, 184–85. Carpenter uses a mean flow at the Yuma gauge of 17.4, not 17.3; million acre-feet per year. Since the gauge record in 1902 was only partial, he used 1903–20.

36. The 6,264,000 acre-feet is 36 percent of the 17.4-million-acre-foot flow at Yuma. Adding in 14 percent (Fall-Davis report, table 1) for the Lower Basin tributaries gives it 50 percent of the available flow.

37. This calculation is spelled out in detail in paragraphs 1 and 2 of his proposed compact; see Minutes of the Eleventh Meeting, 22–23. Further, there is an undated memo in the Carpenter files that appears to lay out the hydrology behind Carpenter's proposal.

38. During the twelfth meeting Carpenter tells Norviel that he is proposing an equal split of the river, but has a difficult time convincing him. See Colorado River Commission, *Minutes and Record of the First Eighteen Sessions*, twelfth meeting, 53–55.

39. Arizona's Norviel understood this problem. He believed the Carpenter proposal gave the Lower Basin only 6.5 million acre-feet of water for future uses. See Colorado River Commission, *Minutes and Record of the First Eighteen Sessions*, twelfth meeting, 55–56.

40. The minutes of the twelfth, thirteenth, fourteenth, and fifteenth meetings show that the commissioners had considerable dialogue among themselves and with Arthur Powell Davis about the level of existing uses on the Gila River and upstream of Lee Ferry. They discuss the differences between the actual flows of the river at Yuma and Laguna and the reconstruction of the natural flows at Lee Ferry. Hoover and the commissioners refer to the "gaugings at Laguna," and Hundley in *Water and the West* refers to "measurements" at Laguna Dam (193), despite the fact that there was no actual gauge to measure the river's flow at Laguna.

41. Carpenter makes note of this difference during the fifteenth meeting during a discussion of the need for a minimum annual flow. See Colorado River Commission, *Minutes and Record of the First Eighteen Sessions*, fifteenth meeting, 31–32.

42. Hundley, *Water and the West*, 192.

43. In WS556, LaRue corrects the flow at Lee Ferry to reflect a 1922 level of development. His corrected flow at Lee Ferry is 800,000 acre-feet per year less than his estimate actual flow (14.4 vs 15.2); see WS556, table 8, page 112. To correct the Laguna flows, we also need to consider the increasing level of development on the Little Colorado River, the Virgin River, and the main stem below the Virgin, about 200,000 acre-feet per year. Arthur Powell Davis's assumption that the natural or virgin flow of the Colorado River at Lee Ferry and Laguna

are about the same is pretty accurate. A hydrologic reconstruction of the river by Reclamation published in the 1940s shows a natural flow at Lee Ferry of 16.27 million acre-feet per year and at Laguna of 16.45 million acre-feet per year, a difference of only about 1 percent.

44. Hoover suggested the 75 million acre-feet every ten years compromise during the seventeenth meeting. Carpenter responded positively, suggesting they consider it in recess. See Colorado River Commission, *Minutes and Record of the First Eighteen Sessions*, seventeenth meeting, 22–23.

45. By our calculation, had Hoover been using a development-corrected flow for Laguna, he might have suggested a ten-year flow of 77 million acre-feet.

46. Colorado River Commission, *Minutes and Record of the First Eighteen Sessions*, eighteenth meeting, 145.

47. Fall-Davis report, 32.

48. The numerical average for measured flows of the Gila at Dome for 1903–20 is 1,063,000 acre-feet per year. The difference between 1.07 and 1.063 million is probably just a rounding error. Dickinson, *Summary*, 235–41.

49. By the mid-1920s, the Gila River was essentially dry below Phoenix in all but the wetter years.

50. Hundley, *Water and the West*, 198.

51. Because the compact only apportioned water to basins, not states, they were unwilling to state that the apportionment was for Arizona in the wording of the compact.

52. Hundley, *Water and the West*, 199.

53. Colorado River Commission, *Minutes and Record of the First Eighteen Sessions*, sixth meeting, table A, 70.

54. Wilbur and Ely, *Hoover Dam Documents*. Hereafter cited as *Hoover Dam Documents*.

55. Hundley, *Water and the West*, 211–12. However, the Fall-Davis report actually included the development of Native American irrigation projects on its list of potential projects. Their combined use, perhaps over 500,000 acre-feet per year, was not insignificant.

56. Hundley, *Water and the West*, 206–8.

57. Colorado River Commission, *Minutes and Record of Sessions Nineteen Thru Twenty-Seven*, twenty-second meeting, 16–17.

58. Hundley, *Water and the West*, 208; Colorado River Commission, *Minutes and Record of Sessions Nineteen Thru Twenty-Seven*, twenty-first meeting, 25.

59. Hundley, *Water and the West*, 208–10.

60. For example, if during the late summer or fall season, there was insufficient flow to meet the needs of the Imperial Irrigation District, then until storage was built Imperial could take legal action to force rights in the Upper Basin with

junior priorities to curtail their uses. Once the 5 million acre-feet of storage was built, they could not under the compact language do so.

61. Hundley, *Water and the West*, 208.

62. Years later, R. I. Meeker, who during the 1922 compact negotiations was Delph Carpenter's engineering advisor, testifying as an irrigation engineer for Arizona, testified that he assisted A. P. Davis with the reconstruction of the natural or virgin flow of the Colorado River at Lee Ferry. Meeker's reconstructed flow of 17.5 million acre-feet per annum would be a critical number used by Carpenter and his fellow commissioners to sell ratification to the state legislatures. See appendix 4, 78–79, Arizona Reply to Defendant Answer, *Arizona v. California*, 373 U.S. 546, 83 S. Ct. 1468, 10 L. Ed. 2d 542 (1963).

Chapter 4

1. Corthell, "Colorado River Problem," 289–92.

2. Corthell, "Colorado River Problem—II," 387–91.

3. Before Hubert Work became secretary of the interior, he was postmaster general. Trani, "Hubert Work," 31–40.

4. David W. Davis served briefly as commissioner. Before that he was an assistant to the secretary of the interior.

5. *Hoover Dam Documents*, A23–A134.

6. *Hoover Dam Documents*, A31–A43. While Hayden initially supported compact ratification, opposition in his home state was fierce.

7. *Hoover Dam Documents*, A23–A29.

8. *Hoover Dam Documents*, A32.

9. For an interesting discussion see Carlson and Boles, "Contrary Views." Carlson discusses the question "should the Upper Basin forgo, as Lower Basin observers presume it must, a portion of its III (a) allocation in order to deliver to the Lower Basin 75 MAF every ten years?" Carlson goes on to suggest that the Upper Basin would have a reasonable chance of prevailing with this argument in the U.S. Supreme Court.

10. *Hoover Dam Documents*, A105.

11. *Hoover Dam Documents*, A34, A38.

12. *Hoover Dam Documents*, A36.

13. *Hoover Dam Documents*, A48. In 1948, during the negotiations of the Upper Colorado River Basin Compact, R. I. Meeker, who was an engineering advisor to Arizona in 1948 but in 1922 was Carpenter's engineering advisor, told the 1948 compact negotiators that he and Davis were charged with developing a "reconstructed" river. They came up with about 18 million acre-feet at Lee Ferry, which is consistent with Davis's answer to Hayden.

14. Based on information published by LaRue in 1925 (WS556), average depletion upstream of Lee Ferry from 1899 to 1920 was 1.634 million acre-feet per year, close to the 1.7 million acre-feet Davis used. This table suggests that had Davis wanted to do so he could have suggested the commissioners use depletion-adjusted flows during the actual compact negotiations.

15. USBR, *Consumptive Uses*. The average Upper Basin irrigation use within the basin for 2006–10 was 2.7 million acre-feet per year and for exports 820,000 acre-feet per year. Note that the 6.15 million acre-feet is the exact figure for irrigation in the four Upper Division states that Davis presented to the commissioners early in the negotiations. At that time they rejected the number as too small. By 1923, they had accepted the number.

16. *Hoover Dam Documents*, A84–A98.

17. *Hoover Dam Documents*, A82–A83.

18. Hundley, *Water and the West*, 221.

19. *Hoover Dam Documents*, A103.

20. *Hoover Dam Documents*, A121–A133 (Caldwell) and 113–20 (Emerson).

21. *Hoover Dam Documents*, A118. In his report, Caldwell makes clear that he viewed the compact apportionments made under Articles III(a) and (b) as limitations. He states that before either basin could exceed their apportionments, all seven states would have to agree as provided for under Article III(f).

22. *Hoover Dam Documents*, A57–A71.

23. Hundley, *Water and the West*, 220–21. Mackey, *Protecting Wyoming's Share*, 136–38.

24. Mackey, *Protecting Wyoming's Share*, 138.

25. W. N. Searcy, "Notes on the Colorado River Compact," memorandum, January 1923.

26. Hundley, *Water and the West*, 224–25.

27. Hundley, *Water and the West*, 225–31. Initially, Congressman Swing also supported a reservation based on storage. However, he ultimately backed off, instead adopting a strategy of tying congressional ratification of the compact to the authorization of storage. This approach proved successful.

28. Hundley, *Water and the West*, 246.

29. Hundley, *Water and the West*, 246–47.

30. Delphus Carpenter, "E. C. LaRue," undated memorandum, Carpenter Papers, Colorado State University Library. Carpenter also discusses the "differences of opinion" and "considerable personal feelings between the engineers," presumably LaRue and Arthur Powell Davis.

31. Hundley, *Water and the West*, 225–31. Moeller, *Phil Swing and Boulder Dam*, 53.

32. *Hoover Dam Documents*, 36.

33. Parsons, "Colorado River." Parsons attributes most of the opposition to the compact in the 1920s to local politics, reflecting primarily the position of the

Salt River Project (SRP) and mining interests. SRP had developed hydroelectric power on the federally built dams in its project and was marketing much of this power to mines under lucrative contracts and did not want any competition from the Boulder Canyon Project.

Chapter 5

1. "New Homes."
2. LaRue is quoting from WS395.
3. *Testimony of E. C. LaRue, Hearings Before the Senate Committee on Irrigation and Reclamation Pursuant to S. Res. 320 with Respect to Proposed Legislation Relating to the Protection and Development of the Colorado River Basin* (1925), 533.
4. Eugene Clyde LaRue Papers, Huntington Library, San Marino, California, introduction by Diane LaRue, March 2008. Hereafter cited as LaRue Papers.
5. Diane LaRue, introduction to LaRue Papers, 3.
6. Boyer and Webb, *Damming Grand Canyon*.
7. The sale of electricity was essential to the political feasibility of the Boulder Canyon Project. Since the California agencies and power companies would be called on to sign power contracts, they had significant sway over the location of the project.
8. The 1916 WS395 covers 1895–1914, whereas the 1925 WS556 covers 1895–1922 or 1923, depending on the gauge.
9. WS556, appendix A, 101.
10. See WS556 tables 6 and 8, 110 and 112.
11. LaRue never presented a natural flow number, but it can be reconstructed from the information he presented. LaRue corrected the flow record by a 2.35-million-acre-foot-per-year 1922 development level, but in years of low flow, he reduced upstream consumptive use by 25 percent to account for reduced physical water availability.
12. Powell, *Report on the Lands of the Arid Region*.
13. He noted that of the twenty-eight-year period when Colorado River flows reconstructed from the gauge record could be compared with Salt Lake inflows that "the greatest variation occurred in 1908, a variation of 29 percent from the curve. In three other years, the variation was more than 20 per cent, but for 16 of the 28-year period the discharge of the Colorado River varied from that given by the curve less than 10 per cent." WS556, 121.
14. WS556, 121.
15. "Colorado River at Lees Ferry, CO," TreeFlow, accessed January 3, 2019, https://www.treeflow.info/content/upper-colorado.

16. It would be interesting to see how LaRue, or more likely his associate, George Holbrook, who co-authored the water supply section of WS556, made the detailed calculations. Did they utilize a tedious least-squares method or a more simple linear relationship to determine the correlation "curve"? Before the advent of computers and handheld calculators in the 1970s, curve fitting to determine correlations was a cumbersome task.

17. WS395, 196.

18. The early 1900s wet cycle began about 1905 or 1906, so 1899–1914 is not nearly as wet as 1903–20.

19. Stabler's results were documented in a memorandum to the secretary of the interior dated March 17, 1924. In this memo, Stabler states that he actually used two methods, one that overstated and one that understated the actual flow during the calibration period (1903–20), so he uses the average of the two.

20. The House Committee on Irrigation held hearings on the protection and development of the Lower Colorado River basin on March 25, 1924. Secretary Work's report includes a report of a committee of engineers appointed by Work. Stabler's March 17, 1924, memo is included in the engineering committee report. One of Reclamation's representatives was E. B. Debler, who would later issue a water supply study critical to the future of Hoover Dam.

21. WS556, 121.

22. Boyer and Webb, *Damming Grand Canyon*, 263.

23. WS556, 122. For his analysis, LaRue did not consider storage.

24. WS556, 123.

25. Because the Upper Basin has actually been delivering 750,000 acre-feet of water for Mexico, making the total ten-year minimum at Lee Ferry 82.5 million acre-feet every ten years, the shortage, or structural deficit is about 1.2 million acre-feet per year.

26. Boyer and Webb, *Damming Grand Canyon*, 263. See also Langbein, "L'Affaire LaRue."

27. *Hearings Before the Senate Committee on Irrigation and Reclamation Pursuant to S. Res. 320 with Respect to Proposed Legislation Relating to the Protection and Development of the Colorado River Basin* (1925).

28. Smith, George Edson Philip, *Discussion*.

29. The letter from LaRue to Hoover on December 28, 1921, is not included in the LaRue Collection at Huntington Library, but there are two letters acknowledging his offer. The first, from Grover to LaRue, dated January 3, 1922, indicates that the USGS was willing to consider his request, but only after he finished a report on the Klamath he was then working on. The second was a "thank you we'll be in touch" letter from Clarence Stetson, Hoover's assistant, dated January 4, 1922.

30. See Hundley, *Water and the West*, 167. Grover consulted with LaRue on these hearings but would not let him attend. Prior to that hearing on March 3, 1922, Grover wrote LaRue asking him to avoid taking public positions that would be contrary to the USGS's Phoenix testimony. Huntington Library LaRue Collection, 1922 correspondence folder.

31. Langbein, "L'Affaire LaRue," 10.

32. Langbein, "L'Affaire LaRue," 12.

33. Langbein, "L'Affaire LaRue," 11.

34. Boyer and Webb, *Damming Grand Canyon*, 265.

35. Dianne LaRue, introduction to LaRue Papers, 6.

36. Delphus Carpenter, "LaRue," undated memorandum, Carpenter Papers, Colorado State University Library.

37. The 1906–2016 average natural flow at Lee Ferry is 14.8 million acre-feet per year. For the period of 1878–2016, using a combination of the natural flow data base and the earlier work by LaRue and Stabler, it is about 14.5 million acre-feet per year. Most, but not all, climate change studies suggest that as temperatures increase, this average will decline, which will be discussed in later chapters.

38. The Colorado River flow data included in WS395, published in 1916, were not that different from Fall-Davis. It wasn't until Water Supply Paper 556 in 1925 that LaRue made convincing arguments concerning the 1875–1904 dry period and the likelihood of future shortages in the basin.

Chapter 6

1. Crawford, "Sibert Report."

2. Hundley, *Water and the West*, 275.

3. "Mighty Colorado."

4. Sarewitz, "How Science Makes," 385–403.

5. U.S. House Committee on Rules, Hearing on H. Res 146 and H.R. 5773, May 2, 1928.

6. *Hoover Dam Documents*, A185.

7. Sibert and Stevens, *Construction of the Panama Canal*.

8. *Hoover Dam Documents*, appendix 302.

9. *Hoover Dam Documents*, A197.

10. *Hoover Dam Documents*, A198.

11. *Hoover Dam Documents*, A200.

12. *Hoover Dam Documents*, appendix 302, A200.

13. Hundley, *Water and the West*, 275.

14. Charles Winter, letter to the editor, *Big Piney Examiner*, February 14, 1929.

Chapter 7

1. Herbert Hoover, "Statement on Signing a Proclamation on the Colorado River Compact and the Boulder Canyon Project," June 25, 1929, American Presidency Project, UC Santa Barbara, https://www.presidency.ucsb.edu/documents/statement-signing-proclamation-the-colorado-river-compact-and-the-boulder-canyon-project.
2. Herbert Hoover, "The President's News Conference," June 25, 1929, American Presidency Project, UC Santa Barbara, https://www.presidency.ucsb.edu/documents/the-presidents-news-conference-793.
3. Hoover, "President's News Conference."
4. Hundley, *Water and the West*, 268–71.
5. *Hoover Dam Documents*, appendix 205, A35.
6. Hundley, *Water and the West*, 277–81.
7. Carlson and Boles, "Contrary Views." The first two general counsels of the Colorado River Water Conservation District that Eric Kuhn worked with, Don Hamburg and Kenneth Balcomb, also made this argument.
8. Hiltzik, *Colossus*.

Chapter 8

1. Chester G. Hanson, "Wilbur Launches Work on Colorado River Dam," *Los Angeles Times*, September 18, 1930.
2. Leveen, "Natural Resource Development."
3. Kelly, "Colorado River Problem."
4. Boulder Canyon Project Act, 45 Stat. 1057 (1928), 43 U.S.C. § 617 et seq.
5. *Hoover Dam Documents*, 63. In 1930 dollars, $207 million is roughly equivalent to $3 billion in 2017 dollars.
6. The Colorado River Natural Flow Database average natural flow at Lee Ferry for 1921–28 is 17.5 million acre-feet per year, 18 percent above the long-term average.
7. The 1928 act required the secretary to market project power at a competitive rate. We use the term *roughly* because there are many factors involved: better inflow hydrology will result in, on average, more power head, which is a benefit beyond just the additional water passing through the power plants, but it could also increase the amount of spill water that doesn't drop through the power plants.
8. *Hoover Dam Documents*, 78.
9. *Hoover Dam Documents*, 439.

10. *Hoover Dam Documents*, 80. Naming the dam after Hoover was not well received. After Hoover lost the presidency to Roosevelt in 1932, the name was restored to Boulder Dam. In 1947, Congress restored Hoover's name to the dam.

11. In December 1934, after the water and power contracts were signed and Hoover Dam was under construction, Debler authored an updated hydrology study that included the early 1930s drought. He concluded the 1897–1934 average natural flow at Lee Ferry was 16.66 million acre-feet. The report showed that if the Upper Basin were to consume 7.5 million acre-feet per year, the long-term average depleted flow at Lee Ferry would be 9.1 million acre-feet per year, similar to LaRue's WS556.

12. Holdren and Turner, "Characteristics of Lake Mead."

13. Wilbur and Ely, *Hoover Dam Power*. Appendix 30 shows 100 percent of the repayment being provided by power revenues.

14. *Hoover Dam Documents*, 102.

15. Wilbur and Ely, *Hoover Dam Power*, appendix 37.

16. Wilbur and Ely, *Hoover Dam Power*, 68–69. The original April 1930 contract for 1,050,000 acre-feet per year was amended to 1,100,000 acre-feet in the September 1931 contract.

17. The purpose of the reservoir behind Parker Dam, Lake Havasu, is primarily to provide a stable water surface (forebay) for the aqueduct's pumps. Lake Havasu was later used as the forebay for the Central Arizona Project.

18. On today's river, this language has led to opportunities for conservation projects such as Brock Reservoir, which provides storage off the All-American Canal to help manage and reduce the amount of water delivered to Imperial Irrigation District that ends up unneeded. However, the language has put the burden of paying for better management and conservation on the federal government and the more junior users.

19. *Hoover Dam Documents*, 108.

20. R. E. Caldwell, Utah's compact commissioner, stated in his compact report that each basin would be responsible for its reservoir evaporation and would be counted against each's Article III apportionments.

21. *Hoover Dam Documents*, 102.

22. *Hoover Dam Documents*, 110.

23. *Hoover Dam Documents*, 111.

24. *Hoover Dam Documents*, 113. Through the contracts with Arizona and Nevada, Arizona agreed to share up to 4 percent of its 50-percent share of the surplus with the State of Nevada. Nevada agreed to pay 50 cents per acre-foot for the water contracted.

25. Colonel William Donovan was a lawyer in private practice at the time of his appointment. During World War I he received a congressional Medal of Honor. He is perhaps best known as the World War II director of the OSS, the predecessor to the CIA.

26. *Hoover Dam Documents*, 101–2.

27. *Hoover Dam Documents*, 102.

Chapter 9

1. NBER Macrohistory Database (for M0892AUSM156SNBR, accessed December 28, 2017), https://fred.stlouisfed.org/series/M0892AUSM156SNBR; Hiltzik, *Colossus*, 178.

2. USBR, Colorado River Natural Flow Database; USGS, *Surface Water Supply of the United States, 1930*; USGS, *Surface Water Supply of the United States, 1931*.

3. Editorial, *New York Times*, March 14, 1931.

4. The math is straightforward. For the ten-year period of 1931 to 1940, the total actual flow at Lee Ferry was 101.7 million acre-feet. During that time, the Upper Basin had consumed 20.3 million acre-feet, making the total natural flow 122 million acre-feet. Deducting the 75-million-acre-foot delivery under III(c) leaves a balance of only 47 million acre-feet for use by the Upper Basin. To use its Article III(a) apportionment of 75 million acre-feet, the Upper Basin would need 28 million acre-feet of storage in place at the beginning of the drought.

5. "C–BT History," Northern Colorado Water Conservancy District, accessed December 16, 2018, http://www.northernwater.org/AboutUs/C-BTHistory.aspx.

6. Tyler, *Last Water Hole in the West*, 350.

7. *Hoover Dam Documents*, 160–62.

8. *Arizona v. California*, 283 U.S. 423, 51 S. Ct. 522, 75 L. Ed. 1154 (1931). Arizona's arguments in this case are revealing. It tried to make the case that the Lower Basin was already consuming 6.5 million acre-feet out of an apportionment of 7.5 million acre-feet and that the remaining 1 million acre-feet was given to California under the MWD contract. To get a total use of 6.5 million acre-feet, Arizona had to assume that use on its tributaries was about 3 million acre-feet. This left no water under Article III(a) for Arizona. In contrast, the Upper Basin was only using 2.5 million acre-feet; therefore, it had 5 million left for use. See *Hoover Dam Documents*, A1780–81.

9. *Arizona v. California*, 298 U.S. 558, 80 L. Ed. 1331, 56 S. Ct. 848 (1936).

10. *Hoover Dam Documents*, A560.

11. *Hoover Dam Documents*, A5634.

Chapter 10

1. The 1940 "natural flow" was 9.9 million acre-feet at Lee Ferry, according to the USBR Colorado River Natural Flow Database.

2. Noble, Perkins, and Nuremberger, *Foreign Relations*, document 1153.

3. Meyers and Noble, "Colorado River."

4. Committee of Fourteen of the Colorado River Basin States and Committee of Sixteen of the Colorado River Basin States, *Proceedings of the Committee of Fourteen and Committee of Sixteen of the Colorado River Basin States.*

5. Committee of Fourteen of the Colorado River Basin States and Committee of Sixteen of the Colorado River Basin States, *Proceedings of the Committee of Fourteen and Committee of Sixteen of the Colorado River Basin States,* 25.

6. Jacob and Stevens presented two scenarios, one based on likely development (9.1 million acre-feet for the Lower Basin and 6 million acre-feet for the Upper Basin), the other strictly based on the compact apportionments (8.5 million acre-feet for the Lower Basin and 7.5 million acre-feet for the Upper Basin).

7. Committee of Fourteen of the Colorado River Basin States and Committee of Sixteen of the Colorado River Basin States, *Proceedings of the Committee of Fourteen and Committee of Sixteen of the Colorado River Basin States,* 28–30.

8. Committee of Fourteen of the Colorado River Basin States and Committee of Sixteen of the Colorado River Basin States, *Proceedings of the Committee of Fourteen and Committee of Sixteen of the Colorado River Basin States,* 1939, 27.

9. Hundley, *Dividing the Waters,* 100.

10. Hundley, *Dividing the Waters,* 101. If the United States had decided to wait until Arizona and California had settled their issues, we would still be waiting.

11. As will be explored in later chapters, the Supreme Court would apportion Lower Basin main stem uses in and below Lake Mead by interpreting the congressional intent in passing the 1928 act. There is no compact among the states with Lower Basin interests and may never be.

12. Hundley, *Dividing the Waters,* 102–4.

13. Committee of Fourteen of the Colorado River Basin States, *Proceedings of the Committee of Fourteen of the Seven States of the Colorado River Basin: Cortez Hotel, El Paso, Texas, June 17–20, 1942.*

14. Committee of Fourteen of the Colorado River Basin States, *Proceedings of the Committee of Fourteen of the Seven States of the Colorado River Basin: Cortez Hotel, El Paso, Texas, June 17–20, 1942,* 98–100. For a good overview of these events see Hundley, *Dividing the Waters,* 107–9. The fact that Utah opposed Tipton's suggestion shows that the Upper Basin states were not always monolithic. Utah's opposition was in part based on Tipton's assumption that the Upper Basin would not fully consume its 7.5 million acre-feet.

15. The decision to do this was to avoid providing Mexico with an incentive to become involved in the internal operations of the river within the United States. The treaty did provide that 500,000 acre-feet be delivered through the Pilot Knob power plant through 1980, then 375,000 acre-feet per year after that.

16. Hundley, *Dividing the Waters*, 134.

17. Tipton, *Statement Concerning Miscellaneous Items*.

18. Hundley, *Dividing the Waters*, chapt. 6.

19. Statement of Lawrence M. Lawson, Hearings on Treaty with Mexico Relating to the Utilization of Waters of Certain Rivers, Before the Committee on Foreign Relations, Seventy-Ninth Congress, 1945.

20. Arthur Powell Davis estimated the natural flow for Yuma was 20.6 million acre-feet per year; Carpenter used 20.5 million acre-feet per year.

21. *Hoover Dam Documents*, 160–62.

22. Smith, *Arizona Loses a Water Supply*.

23. Hundley, *Dividing the Waters*, 169.

Chapter 11

1. Papers of George E. P. Smith, University of Arizona Special Collections Library, https://speccoll.library.arizona.edu/collections/papers-george-e-p-smith.

2. Smith, *Arizona Loses a Water Supply*.

3. George E. P. Smith to O. C. Williams, Arizona State Land Commissioner, October 27, 1944, in Smith, *Arizona Loses a Water Supply*.

4. Smith, *Arizona Loses a Water Supply*, 2. Today based on the Consumptive Uses and Losses Reports prepared by the USBR, the Upper Basin is using about 4.5 million acre-feet per year, including reservoir evaporation; the Lower Basin is using about 10.5 to 11.5 million acre-feet per year if main stem reservoir evaporation and consumptive uses on the Gila River are included.

5. George E. P. Smith to Senator Carl Hayden, January 16, 1945, in Smith, *Arizona Loses a Water Supply*.

6. Smith, *Arizona Loses a Water Supply*, 13.

7. Smith, *Arizona Loses a Water Supply*, 15–16.

8. Smith, *Arizona Loses a Water Supply*, 18.

Chapter 12

1. "Tenth Birthday of Hoover Dam Observed Today," *Los Angeles Times*, October 22, 1946.

2. Reisner, *Cadillac Desert*, 149.

3. USBR, *Colorado River: Natural Menace*, 18. Commonly referred to as House Document 419 or the Blue Book, hereafter cited as HD 419. The original 1946 draft report was included in its entirety in HD 419, along with a lengthy introduction offering each state's formal comments.

4. HD 419, 5.

5. Arizona, which has a small amount of land in the Upper Basin, was also participating in the Upper Colorado River Basin Compact negotiations, which we shall see cemented the close working relationship between it and the four upper states that began during the Senate ratification of the Mexican treaty.

6. The HD 419 hydrology showed that the annual natural flow at Lee Ferry (1897–1943) was 16.27 million acre-feet per year and the flow near Yuma (above the confluence with the Gila) was 16.41 million acre-feet per year. The tributaries between Lee Ferry and Imperial Dam contribute about 900,000–1 million acre-feet; therefore, the net losses are about 800,000–900,000 acre-feet per year. The estimate used by federal officials during the Senate ratification of the Mexican treaty was 400,000 acre-feet per year.

7. The 1928 act limited California to 4.4 million acre-feet per year of Article III(a) water plus one-half of the unapportioned surplus using diversion less return flow as the standard. It also suggested a Lower Basin compact where California would get 4.4, Arizona 2.8, and Nevada 0.3 million acre-feet per year of beneficial consumptive use.

8. Exports out of the basin are fully consumptive. No return flows accrue to the stream of origin.

9. In their 1930 case, Arizona claimed it was using already using 3 million acre-feet of apportioned water. To get a number that large, it had to use the diversion-less-return-flow theory.

10. During the ratification of the 1944 treaty with Mexico, engineering experts representing the United States testified that salvage by use would increase the water supply on the main stem by 380,000 acre-feet per year.

11. Nevada gets 300,000 acre-feet under the 1928 act. Arizona believed 131,000 acre-feet would satisfy Utah's and New Mexico's Lower Basin use in the Virgin River and on the Upper Gila River.

12. The total of three estimates of salvage by use: 1 million for the Gila; 400,000 for the lower main stem; and 600,000 for above Lee Ferry.

13. HD 419, 25.

14. HD 419, 40–42. Under California's math, Arizona would be using 3.57 million acre-feet per year, not including its hoped-for Central Arizona Project.

15. Hundley, *Water and the West*, 209.

16. Wyoming's was the exception. As we shall see in chapter 13, during the negotiations of the 1948 Upper Basin Compact, W. J. Wehrli expressed concerns about

how the stream depletion theory would impact the Upper Basin's obligation to Mexico.

17. The authors believe that based on the most recent twenty-five to thirty years of hydrology and the likely impacts of climate change, the water available on the Colorado River (using either the stream-depletion or the diversions-less-return-flow theories), including the Gila, is significantly less than 16 million acre-feet per year. The Upper Basin still has an argument that because it is only using 4.5 million acre-feet of its III(a) water, the surplus should be calculated as the amount above 8.5 million acre-feet plus the 4.5 million acre-feet the Upper Basin is actually using, or 13 million acre-feet.

18. Hundley, *Dividing the Waters*, 209.

19. Tipton, *Statement Concerning Miscellaneous Items*.

20. Every five years the Bureau of Reclamation publishes a Consumptive Uses and Losses Report for the Colorado River Basin. The last year consumptive use data have been reported for the Gila River is 2005. The 2006–10 report includes the following comment on page 2: "Because of an ongoing effort in the Lower Basin to refine and recalculate in that reach of the river (and its tributaries), this report covers only upper basin data. When the lower basin data [are] available for this five year reporting period, it will be included in a total basin report." As of December 2018, the data were still not available.

21. Webb, *Tree Rings and Telescopes*, 175.

22. Using the average of reconstructed annual flows at Lee Ferry from LaRue and Stabler for the 1878–96 period plus HD 419's Lee Ferry flows for 1897–1943, the natural flow at Lee Ferry is about 15.4 million acre-feet per year.

Chapter 13

1. Memorandum by Delphus Carpenter, "Preliminary Suggestion for an Upper Colorado River Compact," November 1929, Carpenter Papers, Colorado State University Library.

2. Jean S. Breitenstein, Memorandum on the Colorado River to the Colorado Water Conservation Board, August 3, 1947, collection of the Colorado River Water Conservation District.

3. Tipton, *Report on Water Supply*.

4. Tipton, *Report on Water Supply*, 17.

5. Bashore's primary advisor was John R. (Randy) Riter, a hydraulic engineer from the Denver office of the Bureau of Reclamation. Riter was the obvious chair of the engineering committee. He had access to the talented engineering staff, data, and financial resources of his agency.

6.   Like the 1922 compact minutes, the official record of the commission does not include the committee proceedings, executive sessions, off-the-record discussions and the private discussions where most of the horse trading and difficult compromises actually occurred.

7.   One of the reasons for the difference is that the Engineering Committee used a different period of record (1914–45) than the HD 419 hydrology (1897–1943), which was the basis of the Senate testimony on the 1944 treaty. The Engineering Committee picked 1914–45 because it included a sixteen-year wet period (1914–29) and a sixteen-year dry period (1930–45), and they had little confidence in the gauging records before 1914. See Upper Colorado River Basin Compact Commission, *Official Record Vol. III*, 2.

8.   Upper Colorado River Basin Compact Commission, *Official Record Vol. I*, meeting 5, 72.

9.   The other subtle advantage was that the state that would gain the most from salvage by use was Colorado. Stone, Breitenstein, and Tipton all understood that Colorado would be giving up water to the other states. The salvage water might help sell the compact back home.

10. As previously mentioned, during the 1922 compact negotiations, Meeker was Carpenter's primary engineering advisor. As far as we can tell, Meeker is the only individual who had significant roles in the negotiations of both the 1922 and 1948 compacts.

11. Wehrli even made the comment that had the 1922 compact negotiators intended Tipton's interpretation they would have written paragraph III(a) differently.

12. Upper Colorado River Basin Compact Commission, *Official Record Vol. II*, meeting 7, 55–57. With twenty-twenty hindsight, Wehrli and the Wyoming delegation were on point. It's somewhat shocking the commissioners and their advisors showed little interest in the Mexican treaty issue. In a 1945 memo to the Colorado Water Conservation Board, Breitenstein predicted that the dispute over the interpretation of the Mexican treaty provision, Article III(c), was headed to the Supreme Court.

13. The meeting, held in Vernal, Utah, started on July 7, 1948, and concluded on July 21, 1948. Most of that time was spent by subcommittees dealing with engineering, legal, and drafting issues.

14. Upper Colorado River Basin Compact Commission, *Official Record Vol. II*, meeting 7, 63–69. This number is sometimes reported as 117 percent if Arizona's request is included.

15. Upper Colorado River Basin Compact Commission, *Official Record Vol. II*, meeting 7, 99.

16. Upper Colorado River Basin Compact Commission, *Official Record Vol. II*, meeting 7, 116. When the Indian projects in New Mexico and Utah were added

to the Reclamation number, the total potential depletions were 9.446 million acre-feet.

17. It was a long and perhaps contentious caucus. Utah asked for two delays waiting for members of its delegation. Upper Colorado River Basin Compact Commission, *Official Record Vol. II*, 123.

18. John Fleck, "Water Tug of War Goes On," *Albuquerque Journal*, February 10, 2013.

19. *Winters v. United States*, 207 U.S. 564, 28 S. Ct. 207, 52 L. Ed. 340 (1908).

20. The 1948 compact negotiators were faced with what we might now call a catch-22. If the apportionments were to cover all potential uses then New Mexico's apportionment had to be large enough so that it would cover the collective amount of water required to satisfy rights that would be unimpaired by the compact.

21. The commission minutes are silent on why Wyoming gave up on its opposition. We suspect that they gave it up in return for Colorado dropping its request down to 51.75 percent.

22. The inflow-outflow method can roughly be explained as measuring the total water flowing into the top of the river system, then subtracting the amount flowing out the bottom. The difference is the net consumptive use. In reality, it is much more complicated.

23. Article III(b)(3) of the 1948 compact actually states, "No State shall exceed its apportioned use in any water year when the effect of such use, as determined by the commission, is to deprive another signatory State of its apportioned use during that water year." The most likely way this could occur is if a state's overuse were to result in a curtailment.

24. The amount of water a state owed is the deficiency (as determined by the commission) times its percentage of the total postcompact uses in the Upper Division states. Article IV of the 1948 compact is confusing, and there isn't a consensus among the Upper Basin states as to how it would ever be implemented.

25. The 1948 compact refers to Article III in its entirety, not just Article III(c), the 75-million-acre-feet-every-ten-years provision.

26. This debate was expected. In fact, the 1948 compact states took the position that they were convening under the general authority of Article 10 of the constitution and not Section 14 of the 1928 act. They feared that if they used the act's authority, California would make a legal claim that it needed to be a party to an Upper Basin compact.

27. The Californians probably wanted more than just report language, but they were boxed in. The subcommittee chair, John R. Murdock, was from Arizona. See *Upper Colorado River Basin Compact*, Hearing before the House Subcommittee Irrigation and Reclamation of the Committee on Public Lands, 81 Cong. (1949) H.R. 2325, 81 H.R. 2326, 81 H.R. 2327, 81 H.R. 2328, 81 H.R. 2329, 81 H.R.

2330, 81 H.R. 2331, 81 H.R. 2332, 81 H.R. 2333, 81 H.R. 2334. The subcommittee then unanimously recommended congressional consent.

Chapter 14

1. Hearing on H.R. 3383 before the House Interior Committee, 84th Congress, 1st Session, 333.
2. 100 Cong. Rec. A6023 (statement of Senator John Saylor, quoting Jacobson).
3. 43 U.S.C. § 620. The act also is sometimes referred to as the Colorado River Storage Project and Participating Projects Act.
4. For example, an irrigation district paying off $50 million over fifty years with no interest would pay $1 million per year. At an annual interest rate of 4 percent, it would pay $2.3 million per year, a difference of $65 million over the repayment period.
5. The 1956 act identified projects that were authorized immediately, such as the Central Utah Project, and projects that were to be studied for future authorization.
6. Nevada was an ally of California on the Mexican treaty but supported the CRSPA.
7. Blue Mesa, Morrow Point, and Crystal Dams. Blue Mesa is the main storage reservoir, Morrow Point is primarily for power generation, and Crystal is a reregulating reservoir that allows the hydroelectric power plants installed at Morrow Point and Blue Mesa to be used for load following or peaking purposes. Its storage capacity is used to reregulate the flows from Morrow Point, which are large during the afternoons and early evenings when power demand peaks and small at night.
8. The term *cash register dam* first entered widespread use in the 1960s in discussions over the need for construction of a Colorado River dam to power Central Arizona Project pumps and fund augmentation. But the concept goes back further and was central to the economics of the construction and operation of Hoover Dam.
9. After consultation with the basin states, the secretary of the interior issued formal filling criteria in 1962.
10. The irrigation features of the Animas-La Plata (Colorado) and the Seedskadee (Wyoming) projects were eliminated, but the storage features were completed.
11. Hurley, *United States Census of Agriculture, 1954*; U.S. Department of Commerce, *1982 Census of Agriculture*.

Chapter 15

1. Clyde, "Present Conflicts"; MacDonnell, "*Arizona v. California*: Its Meaning."
2. *Hoover Dam Documents*, A35.

3. Motion for Leave to File Bill of Complaint, *Arizona v. California*, 373 U.S. 546 (1963).

4. As of 1952, the CAP was contemplated to divert about 1.2 million acre-feet per year. During the debate over its authorization, it was upsized to its current capacity of about 1.6 million acre-feet per year.

5. The 3.8-million-acre-foot number is California's basic apportionment of 4.4 million acre-feet less about 600,000 acre-feet of main stem reservoir evaporation that would be charged to California under Arizona's claims.

6. Answer of Defendants to Bill of Complaint, *Arizona v. California*, 373 U.S. 546 (1963).

7. Recall that under E. B. Debler's 1930 Hoover Dam hydrology, the average inflow to Lake Mead was 11.9 million acre-feet per year, but the water available from Lake Mead for downstream deliveries was only 10.5 million acre-feet per year. The 1.4-million-acre-foot difference was to account for evaporation and system losses.

8. Examples are the legal challenges by environmental groups to the secretary of the interior's decision to line the All-American Canal and the designation of critical habitat for the Southwestern Willow Flycatcher under the Endangered Species Act.

9. Exceptions of the California Defendants to the Report and Recommendation of the Special Master with Respect to Their Motion to Join the States of Colorado, New Mexico, Utah, and Wyoming, *Arizona v. California*, 373 U.S. 546 (1963), 63.

10. Brief of the State of Colorado and the State of Wyoming Opposing the Motion of California to Join the States of Colorado and Wyoming as Parties to This Action, *Arizona v. California*, 373 U.S. 546 (1963), 37.

11. Special Master's Report on the Motion of the California Defendants to Join as Parties the States of New Mexico, Utah, Colorado, and Wyoming, *Arizona v. California*, 373 U.S. 546 (1963).

12. In one of his more interesting conclusions, Haight argues that how consumptive use is to be measured "is a question for each Basin to determine so long as the Colorado Compact endures." Our reading of Haight's report is that he had little understanding of how the future use of the surplus for Mexico or for reapportionment under Article III(f) tied the two basins together.

13. Special Master's Report on the Motion of the California Defendants to Join as Parties the States of New Mexico, Utah, Colorado, and Wyoming, *Arizona v. California*, 373 U.S. 546 (1963).

14. Special Master's Report on the Motion of the California Defendants to Join as Parties the States of New Mexico, Utah, Colorado, and Wyoming, *Arizona v. California*, 373 U.S. 546 (1963).

15. *Arizona v. California*, 350 U.S. 114 (1955).

16. August, *Dividing Western Waters*, chapt. 5.

17. The 1964 decree refers to these rights as "present perfected rights" and defines them as rights perfected by June 25, 1929, the date the Boulder Canyon Project Act took effect.

18. In theory, the Consumptive Uses and Losses Report prepared by the secretary of the interior pursuant to the 1968 Colorado River Basin Project shows the amount of consumptive uses on the Lower Basin tributaries, but the data on the Gila was last updated in 2005 and includes a footnote that consumptive uses on the Gila are almost impossible to determine.

19. It was a compelling argument because of the math of the 1930s and 1953–64 droughts. In order to meet its obligations under Articles III(c) and (d), the Upper Basin needed 25–30 million acre-feet of storage upstream of Lee Ferry. Although it opposed the 1956 act based on the subsidies provided to the irrigation projects, even California in its comments to HD 419 acknowledged the need for significant storage in the Upper Basin. An argument that supports the Upper Basin's decision to stay out of the case is that during the mid to late 1960s two major nationwide movements were beginning to have a major impact on Congress. The first was that Congress was becoming far more fiscally conservative. The idea that the federal government needed to subsidize agricultural water development in order to encourage the settlement of the West was outdated. For many reasons, people were migrating to the West. The second was the beginnings of the environmental movement. For both of these reasons, obtaining congressional approval of large reservoirs with major environmental footprints and traditional irrigation projects with large subsidies would have been more difficult in the 1960s than it was in the 1950s.

20. Report from Special Master Simon H. Rifkind, *Arizona v. California*, 373 U.S. 546 (1963). See 146–49. Rifkind clearly favors California's diversions-less-return-flows theory. However, he also makes clear that resolution of the contested issues related to the compact is not necessary for disposition of the case in front of him.

21. If one assumes the diversions-less-return-flow theory of accounting, Arizona would be using about 2 million acre-feet of water on the Gila, 1.2 million acre-feet of main stem use, and about 400,000 acre-feet of its share of main stem reservoir evaporation. Even if the decision had awarded Arizona the 3.8 million acre-feet it was asking for (unlikely), this would leave only 200,000 acre-feet available for a CAP.

22. Assuming 1 million acre-feet per year of evaporation and system losses on the main stem, if California's share was 4.4/7.5, 586,666 acre-feet would be charged

to it. Even if California had received 50 percent of the III(b) water, it would still be a net loss of 86,666 acre-feet.

23.   MacDonnell, "*Arizona v. California* Revisited."

Chapter 16

1.   Letter from the Secretary of the Interior Transmitting a Report and Findings of the Central Arizona Project, H.R. Doc. No. 136, 81st Cong., 1st Sess. (1949).

2.   Hearing on H.R. 4671 Before the House Subcommittee on Irrigation and Reclamation, 89th Congress (1965) (statement of Floyd Dominy).

3.   Hearing on H.R. 4671 before the House Subcommittee on Irrigation and Reclamation, 89th Congress (1965) (statement of Floyd Dominy).

4.   During the three-year congressional debate over the CAP, the period of record used by the Bureau of Reclamation changed. By early 1968, the period of record was 1906–67. Hearings on H.R. 3300 and S. 1004 before the Subcommittee on Irrigation and Reclamation, 90th Congress (1968).

5.   Upper Colorado River Commission, *Water Supplies of the Colorado River.*

6.   The other committee members were W. E. Steiner and D. E. Cole from California, I. P. Head from Nevada, and W. S. Gookin from Arizona.

7.   Hearing on H.R. 4671 before the House Subcommittee on Irrigation and Reclamation, 89th Congress (1965) (statement of W. Don Maughan).

8.   The USGS was a catalyst to these advancements. See Leopold, *Probability Analysis Applied to a Water-Supply Problem.*

9.   Johnson, *Central Arizona Project.*

10. Colorado River Basin Project, Hearing on H.R. 330 and S. 1004, House Subcommittee on Irrigation and Reclamation, 90th Cong. (1968) (statement of Floyd Dominy).

11. Colorado River Basin Project, Hearing on H.R. 330 and S. 1004, House Subcommittee on Irrigation and Reclamation, 90th Cong. (1968).

12. The Upper Basin's estimates of its requirements without regard to water availability was 7.35 million acre-feet per year. Tipton in his 1965 study assumed that based on only a 75-million-acre-foot delivery, the Upper Basin would be limited to 6.3 million acre-feet per year.

13. The 5.43-million-acre-foot number was for the year 2000, for the year 2013 Reclamation assumed the Upper Basin depletions would reach a level of 5.8 million acre-feet annually.

14. Hearing on H.R. 4671 Before the House Subcommittee on Irrigation and Reclamation, 89th Congress (1965). Page 539, table 3, includes a note that the Lower Basin engineering committee (Maughan Committee) used a number of 5.5 million acre-feet per year for its analyses.

15. Hearing on H.R. 4671 before the House Subcommittee on Irrigation and Reclamation, 89th Congress (1965) (statement of Stewart Udall).

16. In the 1960s, during the debate over the Grand Canyon, the boundaries of the Grand Canyon National Park were smaller than today's park.

17. The 2.1-billion-kilowatt-hour estimate was based on a delivery of 1.2 million acre-feet per year. As built, the CAP delivers about 1.6 million acre-feet per year, so today's annual energy needs are about 2.8 billion kilowatt hours.

18. Reisner, *Cadillac Desert*, 134.

19. Hearings Before the Senate Committee on Water and Power Resources on S. 1004, 90th Cong. (1967).

20. Hearings Before the Senate Committee on Water and Power Resources on S. 1004, 90th Cong. (1967). The 822,000 acre-feet is the amount of water available without capacity limitations. If the project was limited by a 2,500 cfs canal, the 2030 average was 676,000 acre-feet per year.

21. Lyndon B. Johnson, "Remarks Upon Signing the Colorado River Basin Project Act," September 30, 1968, American Presidency Project, U.C. Santa Barbara, https://www.presidency.ucsb.edu/documents/remarks-upon-signing-the-colorado-river-basin-project-act.

22. Udall, Neuman, and Udall, *Too Funny to Be President*, 61.

23. 43 U.S.C. § 1501 et seq. Section 602(a) specifies that during wet years releases are also made to avoid physical spills at Glen Canyon. During wet cycles such as 1983–86 or 1995–98, the operations of the reservoir are controlled by the avoidance of spills.

24. 43 U.S.C. § 1501 et seq. Section 601 requires the secretary to promulgate the LROC, then formally review it every five years.

25. The reasons the current Colorado River Natural Flow Database starts in 1906 are a mystery. The best speculation is that since river models required the flows to be disaggregated by tributaries and into monthly flows, there was not enough gauge data to have any confidence in the disaggregation. Personal communication with James Prairie, USBR.

26. Eric Kuhn's predecessor at the Colorado River District, Rolly Fischer, participated on the task force and kept detailed notes. Both Fischer and then CWCB director Felix Sparks told the story that Secretary Hickel had made the decision early to release 8.23 million acre-feet annually from Glen Canyon and the task force's real purpose was to justify it.

27. The release goal is formally called a "minimum objective release," language carefully chosen to suggest it is a goal—an "objective"—but not a legal requirement. While creating a practical delivery obligation of 8.23 million acre-feet that has been honored operationally ever since, it avoided the legal dispute

about whether that delivery—specifically the Upper Basin's share of the Mexican treaty obligation—was a legal requirement.

Chapter 17

1. Berkman and Viscusi, *Damming the West.*
2. Remarks by William Lane and Mike Clinton to the Colorado River Water Users Association, December 1977, files of the Colorado River Water Conservation District.
3. Webb, *Tree Rings and Telescopes.*
4. "Pioneering Work in the Colorado River Basin (1940s)," TreeFlow, accessed May 22, 2018, https://www.treeflow.info/content/pioneering-work-colorado -river-basin-1940s.
5. Lane, "Potential Use." One reason progress slowed may have been the untimely death of Schulman at the age of 49.
6. Lane, "Potential Use."
7. Stockton and Jacoby, "Long-Term Surface-Water Supply."
8. The Bureau of Reclamation's Colorado River Natural Flow Database starts in 1906. The latest tree-ring-based reconstruction shows that the 1906–30 period as the wettest twenty-five-year period since 1400 ("Colorado River at Lees Ferry, CO," TreeFlow, accessed April 14, 2019, https://www.treeflow.info /content/upper-colorado).
9. Young, "Coping."
10. More than two decades after the study was published there has been no serious consideration of replacing the 1922 compact, but some of the study recommendations, such as providing for short term shortage sharing and restoring the river delta, are underway.
11. "Colorado River at Lees Ferry, CO," TreeFlow, accessed April 14, 2019, https:// www.treeflow.info/content/upper-colorado.
12. Reisner and Bates, *Overtapped Oasis.*
13. Philip Shabecoff, "Haste of Global Warming Trend Opposed," *New York Times,* October 21, 1983.
14. Revelle and Waggoner, "Effects."
15. Stockton and Boggess, *Geohydrological Implications.*
16. Peterson, Connolley, and Fleck, "Myth."
17. National Research Council, *Changing Climate.*
18. Kiparsky and Gleick, *Climate Change.*
19. National Research Council, *Colorado River Basin,* 91.
20. National Research Council, *Colorado River Basin,* 92.

21. Christensen et al., "Effects."
22. USBR, *Colorado River Interim Surplus*, B-46.
23. USBR, *Final Environmental Impact Statement*, U-2.

Chapter 18

1. USBR, "Lake Mead Annual High and Low Elevations (1935–2018)," accessed February 26, 2018, https://www.usbr.gov/lc/region/g4000/lakemead_line.pdf.
2. U.S. Department of the Interior, *Compilation of Records.*
3. On October 17, 2010, Lake Mead dropped to a surface elevation of 1,083.10 feet above sea level, surpassing the lowest levels of the drought of the 1950s. John Fleck, "Drought, American Style," *Jfleck at Inkstain* (blog), October 17, 2010, http://www.inkstain.net/fleck/2010/10/drought-american-style/.
4. Bill Boyarsky, "Sharing the Colorado River's Water: West Braces for a Change," *Los Angeles Times*, September 23, 1985.
5. FRED Economic Data (for AZMARI3POP, accessed February 26, 2018), https://fred.stlouisfed.org/series/AZMARI3POP.
6. LaBianca, "Arizona Water Bank."
7. Because of the several-day travel time from Lake Mead to Parker Dam, the Palo Verde Diversion Dam, and the Imperial Diversion Dam, the delivery of water to U.S. users is not very efficient, often resulting in extra water to Mexico. So even though the treaty delivery is 1.5 million acre-feet, actual deliveries averaged about 1.7 million acre-feet per year. This problem has been reduced by the construction of Brock Reservoir.
8. USBR, Colorado River Natural Flow Database.
9. An additional complicating problem is that under the seven-party agreement, the 3.85 million acre-feet was collectively available to the four agricultural districts, but their individual contract allotments were not quantified, thus before MWD could begin the transfer program, the parties needed to negotiate a quantification settlement agreement, the QSA.
10. Bureau of Reclamation, Department of the Interior, Notice of Availability of Record of Decision for the Adoption of Colorado River Interim Surplus Guidelines, 66 FR 7772 (January 25, 2001).
11. Prior to 2000, there were only seven individual years in the NFDB that had natural flows below 9.4 million acre-feet.
12. Letter from the Colorado, New Mexico, Utah, and Wyoming governors' representatives on Colorado River Operations, April 18, 2005. Precipitation in the Lower Basin was quite abundant during the odd winter of 2005, enough that runoff in the Gila River system actually made it to Yuma and contributed to the U.S. deliveries to Mexico.

13. Letter from Secretary Norton to Governor Jon Huntsman of Utah, May 5, 2005.

14. Fleck, *Water is for Fighting Over.*

15. USBR, *Final Environmental Impact Statement.*

16. Under the guidelines, if Lake Mead is projected to fall below 1,025 feet, reconsultation is triggered. Under this reconsultation, shortages could actually be much greater than 600,000 acre-feet.

17. The perceived problem with the 1970 LROC operation was that at the beginning of a drought, Lake Powell dropped much faster than Lake Mead, but then, when conditions turned wetter, Powell began to recover quickly, but Lake Mead continued to decline.

18. Depending on the modeling used, there were a few more 9-million-acre-foot releases, but during drier periods more 7.48-million-acre-foot releases, which the States of the Upper Division considered a benefit.

Chapter 19

1. John Fleck, "A Chat with Mike Connor," *Albuquerque Journal*, September 18, 2009.

2. U.S. Department of the Interior, *Colorado River Basin Water Supply and Demand Study.*

3. For Lake Powell it is elevation 3,390 feet; for Lake Mead it is elevation 895 feet. At the dead pools there is still some water in each reservoir. It just can't be released—except that with Las Vegas's new tap into the bottom of Lake Mead, once the pumps are installed, it will be able to divert dead pool water.

4. For example, examining model output from the 2007 environmental statement, appendix U, there were several periods during the 1600s where both reservoirs would be completely drained.

5. Woodhouse et al., "Increasing Influence"; Udall and Overpeck, "Twenty-First Century."

6. John Fleck, "New USBR Modeling Suggests a Bigger Risk of Colorado River Shortage Than Y'all Might Think," *Jfleck at Inkstain* (blog), June 23, 2018, http://www.inkstain.net/fleck/2018/06/new-usbr-modeling-suggests-a-bigger-risk-of-colorado-river-shortage-than-yall-might-think/.

Chapter 20

1. One of the authors, Eric Kuhn, started using the stress test along with scientist John Carron of Hydros Consulting for risk studies conducted for the Upper Colorado River Commission Engineering Committee. Kuhn first publicly proposed its use at the 2013 meeting of the Colorado River Water Users

Association—during the same panel discussion in which Arizona's Tom McCann described the "structural deficit."

2. While the 1988–2018 average natural flow at Lee Ferry is 13.1 million acre-feet per year, for the nineteen-year period of 2000–2018, it's been about 12.4 million acre-feet.

3. National Research Council and Carbon Dioxide Assessment Committee, *Changing Climate.*

4. Woodhouse et al., "Increasing Influence"; Udall and Overpeck, "Twenty-First Century"; McCabe et al., "Evidence That Recent Warming Is Reducing Upper Colorado River Flows."

5. Dieter et al., *Estimated Use of Water in the United States in 2015*; Fleck, *Water is for Fighting Over.*

6. Metropolitan Water District of Southern California, annual water use dataset, 1980–2016.

7. U.S. Department of the Interior, *Colorado River Accounting and Water Use Report.*

8. Fleck, *Water is for Fighting Over.*

9. Fleck, "What Seven States Can Agree to Do."

10. The possibility of a future without a compact is based on the theory that the court might conclude that there was a mutual mistake of fact, the hydrology, and throw it out. The additional federalization is based on the concept that one mechanism the court could adopt if faced with major differences between the states is to give secretaries of the Interior far more power than they have today. It certainly did so in the 1964 decree in *Arizona v. California.*

11. The question of how much water is being consumed in the Lower Basin is controversial. The last data available from the Consumptive Uses and Losses Report is 2005. To get to a range of 10.0–11.5 million acre-feet per year, we added 7.5 million acre-feet per year of normal deliveries from Lake Mead, 1–1.5 million acre-feet per year of evaporation and system losses, and 1.5–2.5 million acre-feet per year of Lower Basin tributary use on the Little Colorado, Virgin, Bill Williams, and Gila Rivers. In recent years, the Lower Basin's main stem uses have approached 7 million acre-feet per year. This number assumes the diversions-minus-return-flows system of accounting because the Consumptive Uses and Losses Report uses that methodology. We are not aware of any report that shows basin depletions under the stream depletion theory of accounting.

12. U.S. Department of the Interior, *Upper Colorado River Basin Consumptive Uses and Losses Report.*

13. Some of this additional use will be associated with existing projects that are not yet being fully utilized, like Denver's Dillon Reservoir/Roberts Tunnel system.

There are also a few projects that have received permits and are under construction, such as the Windy Gap Firming Project. The largest new project being permitted is Utah's Lake Powell Pipeline. New consumptive uses will be offset by several factors, including the urbanization of western Colorado communities on irrigated lands and a slow but certain retirement of a number of coal-fueled thermal electric plants. These coal plants, many of them close to the end of their design lives, are currently consuming about 160,000 acre-feet per year in the Upper Basin.

14. Kenney, "Rethinking the Future of the Colorado River."
15. Kuhn, *Risk Management Strategies.*
16. Fleck, *Water is for Fighting Over.*

Epilogue

1. Powell, *Exploration of the Colorado River.*
2. WS395, 167.
3. Lepore, *These Truths*, xvi.
4. *Hoover Dam Documents*, A200.
5. Hundley, *Water and the West*, 352.
6. *Hoover Dam Documents*, A200.

BIBLIOGRAPHY

Books

August, Jack L. *Dividing Western Waters: Mark Wilmer and* Arizona v.California. Forth Worth: Texas A&M University Press, 2007.

Berkman, Richard Lyle, and W. Kip Viscusi. *Damming the West: The Nader Task Force Report on the Bureau of Reclamation*. Washington, D.C.: Center for Study of Responsive Law, 1971.

Boyer, Diane E., and Robert E. Webb. *Damming Grand Canyon: The 1923 USGS Colorado River Expedition*. Logan: Utah State University Press, 2007.

Clarke, Robin T., and Jannet King. *The Water Atlas*. New York: New Press, 2004.

Fleck, John. *Water is for Fighting Over and Other Myths About Water in the West*. Washington, D.C.: Island Press, 2016.

Hiltzik, Michael. *Colossus: Hoover Dam and the Making of the American Century*. New York: Free Press, 2010.

Hundley, Norris. *Dividing the Waters: A Century of Controversy Between the United States and Mexico*. Berkeley: University of California Press, 1966.

Hundley, Norris. *Water and the West: The Colorado River Compact and the Politics of Water in the American West*. Berkeley: University of California Press, 2009.

Johnson, Rich. *The Central Arizona Project, 1918–1968*. Tucson: University of Arizona Press, 1977.

Kahrl, William L. *Water and Power: The Conflict over Los Angeles Water Supply in the Owens Valley*. Berkeley: University of California Press, 1983.

Kiparsky, Michael, Peter H. Gleick, and Pacific Institute for Studies in Development, Environment, and Security. *Climate Change and California Water Resources: A Survey and Summary of the Literature*. Oakland, Calif.: Pacific Institute for Studies in Development, Environment, and Security, 2003.

Leopold, Luna. *Probability Analysis Applied to a Water-Supply Problem*. Washington, D.C.: U.S. Geological Survey, 1959.

Lepore, Jill. *These Truths: A History of the United States*. New York: W. W.Norton, 2018.

Mackey, Mike. *Protecting Wyoming's Share: Frank Emerson and the Colorado River Compact*. Sheridan, Wyo.: Western History Publications, 2013.

Moeller, Beverley Bowen. *Phil Swing and Boulder Dam*. Berkeley: University of California Press, 1971.

National Research Council. *Colorado River Basin Water Management: Evaluating and Adjusting to Hydroclimatic Variability*. Washington, D.C.: National Academies Press, 2007.

National Research Council and Carbon Dioxide Assessment Committee. *Changing Climate: Report of the Carbon Dioxide Assessment Committee*. Washington, D.C.: National Academies, 1983.

Noble, G. Bernard, E. R. Perkins, and Gustave A. Nuremberger, eds. *Foreign Relations of the United States, Diplomatic Papers, 1940*. Vol. 5, *The American Republics*, edited by Victor J. Farrar, Richard B. McCornack, and Almon R. Wright. Washington, D.C.: Government Printing Office, 1961. https://history.state.gov/historicaldocuments/frus1940v05.

Powell, John Wesley. *The Exploration of the Colorado River and Its Canyons*. New York: Dover, 1961.

Powell, John Wesley, et al. *Report on the Lands of the Arid Region of the United States: With a More Detailed Account of the Lands of Utah*. Washington, D.C.: Government Printing Office, 1879.

Reisner, Marc. *Cadillac Desert: The American West and Its Disappearing Water*. New York: Viking, 1986.

Reisner, Marc, and Sarah F. Bates. *Overtapped Oasis: Reform or Revolution for Western Water*. Washington, D.C.: Island Press, 1990.

Ross, John F. *The Promise of the Grand Canyon: John Wesley Powell's Perilous Journey and His Vision for the American West*. New York: Viking, 2018.

Sibert, William Luther, and John Frank Stevens. *The Construction of the Panama Canal*. New York: Appleton, 1915.

Simonds, William Joe. *The Boulder Canyon Project: Hoover Dam*. Denver, Colo.: Bureau of Reclamation History Program, 1995.

Smith, George E. P. *Arizona Loses a Water Supply*. Tucson, Ariz.: Self-published, 1956.

Summit, April R. *Contested Waters: An Environmental History of the Colorado River*. Boulder: University Press of Colorado, 2012.

Tyler, Daniel. *The Last Water Hole in the West: The Colorado–Big Thompson Project and the Northern Colorado Water Conservancy District.* Boulder: University Press of Colorado, 1992.

Tyler, Daniel. *Silver Fox of the Rockies: Delphus E. Carpenter and Western Water Compacts.* Norman: University of Oklahoma Press, 2003.

Udall, Morris K., Bob Neuman, and Randy Udall. *Too Funny to Be President.* Tucson: University of Arizona Press, 2001.

Verburg, Katherine Ott. *The Colorado River Documents, 2008.* Denver, Colo.: U.S. Dept. of the Interior, Bureau of Reclamation, Lower Colorado Region, 2010.

Webb, George Ernest. *Tree Rings and Telescopes: The Scientific Career of A. E. Douglass.* Tucson: University of Arizona Press, 1983.

Government Reports

Colorado River Commission. *Minutes and Record of Sessions Nineteen Thru Twenty-Seven of the Colorado River Commission Negotiating the Colorado River Compact of 1922.* Accessed April 17, 2019. https://wwa.colorado.edu/resources/colorado-river/compact.html.

Colorado River Commission. *Minutes and Record of the First Eighteen Sessions of the Colorado River Commission Negotiating the Colorado River Compact of 1922.* Accessed April 17, 2019. https://wwa.colorado.edu/resources/colorado-river/compact.html.

Committee of Fourteen of the Colorado River Basin States. *Proceedings of the Committee of Fourteen of the Seven States of the Colorado River Basin: Cortez Hotel, El Paso, Texas, June 17–20, 1942.*

Committee of Fourteen of the Colorado River Basin States and Committee of Sixteen of the Colorado River Basin States. *Proceedings of the Committee of Fourteen and Committee of Sixteen of the Colorado River Basin States: Meeting of Committee of Fourteen, March 16, 1939, Denver, Colorado, Meeting of Committee of Fourteen, Nov. 17–18, 1939, Denver, Colorado, Meeting of Committees of Fourteen and Sixteen, Nov. 16, 1940, Boulder City, Nevada.*

Davis, Arthur P. *Eighteenth Annual Report of the Reclamation Service, 1918–1919.* Washington, D.C.: U.S. Department of the Interior, 1919.

Davis, Arthur P. *Irrigation near Phoenix, Arizona.* USGS Water Supply and Irrigation Paper, No. 2. Washington, D.C.: Government Printing Office, 1897.

Dickinson, William Eugene. *Summary of Records of Surface Waters at Base Stations in Colorado River Basin, 1891–1938.* Water Supply Paper No. 918. Washington, D.C.: Government Printing Office, 1944.

Dieter, Cheryl A., Molly A. Maupin, Rodney R. Caldwell, Melissa A. Harris, Tamara I. Ivahnenko, John K. Lovelace, Nancy L. Barber, and Kristin S. Linsey. *Estimated*

Use of Water in the United States in 2015. Reston, Va.: USGS, 2018. https://doi.org/10.3133/cir1441.

Hurley, Ray. *United States Census of Agriculture: 1954, Volume 1: Counties and State Economic Areas, Part 29: Wyoming and Colorado*. Washington, D.C.: Government Printing Office, 1956.

Ives, Joseph C. *Report upon the Colorado River of the West: Explored in 1857 and 1858*. No. 90. Washington, D.C.: Government Printing Office, 1861.

Kenney, Doug. *Rethinking the Future of the Colorado River: Draft Interim Report of the Colorado River Governance Initiative*. Boulder, Colo.: Natural Resources Law Center, Western Water Policy Program, 2010.

Kuhn, Eric. *The Colorado River: The Story of a Quest for Certainty on a Diminishing River*. Glenwood Springs, Colo.: Colorado River Water Conservation District, 2007.

Kuhn, Eric. *Risk Management Strategies for the Upper Colorado River Basin*. Glenwood Springs, Colo.: Colorado River Water Conservation District, 2012.

Lane, William. *Potential Use of Tree-Ring Data in Augmenting Hydrologic Records*. Denver, Colo.: U.S. Bureau of Reclamation, 1976.

LaRue, Eugene Clyde. *Colorado River and Its Utilization*. USGS Water Supply Paper No. 395. Washington, D.C.: Government Printing Office, 1916.

LaRue, Eugene Clyde. *Water Power and Flood Control of Colorado River Below Green River, Utah*. USGS Water Supply Paper No. 556. Washington, D.C.: Government Printing Office, 1925.

Mead, Elwood, et al. *Report of the All-American Canal Board*. Washington, D.C.: Government Printing Office, 1920.

Newell, Frederick Haynes. *Proceedings of First Conference of Engineers of the Reclamation Service, with Accompanying Papers*. No. 93. Washington, D.C.: Government Printing Office, 1904.

Stockton, Charles W., and William R. Boggess. *Geohydrological Implications of Climate Change on Water Resource Development*. Tucson, Ariz.: Stockton and Associates, 1979.

Tipton, R. J. *Report on Water Supply of Colorado River and Allied Matters, for the Upper Colorado River Basin Committee*. Denver, Colo.: Colorado River Water Conservation District Archives, 1938.

Tipton, R. J. *Statement Concerning Miscellaneous Items Relating to the Colorado River*. Consulting Engineer for the Colorado Water Conservation Board, January 1945.

Upper Colorado River Commission. *Water Supplies of the Colorado River, Available for Use by the States of the Upper Division and for Use from the Main Stem by the States of Arizona, California and Nevada in the Lower Basin*. Denver, Colo.: Tipton and Kalmbach, 1965.

Upper Colorado River Basin Compact Commission. *Official Record, Vol. I: Record of Preliminary Governors' Meeting, and for Commission Meetings No. 1–5, Inclusive.* Denver, Colo.: The Commission, 1948.

Upper Colorado River Basin Compact Commission. *Official Record, Vol. II: Record of Commission Meetings No. 6–11, Inclusive.* Denver, Colo.: The Commission, 1948.

Upper Colorado River Basin Compact Commission. *Official Record, Vol. III: Final Draft of Engineering Advisory Committee Report and Inflow-Outflow Manual.* Denver, Colo.: The Commission, 1948.

USBR. "Colorado River Basin Natural Flow and Salt Data." Colorado River Basin Natural Flow Database. Last modified April 1, 2019. https://www.usbr.gov/lc/region/g4000/NaturalFlow/current.html.

USBR. *Colorado River Interim Surplus Criteria Final Environmental Impact Statement.* Washington, D.C.: Department of the Interior, 2000.

USBR. *The Colorado River: Natural Menace Becomes a National Resource: Interim Report on the Status of the Investigations Authorized to Be Made by the Boulder Canyon Project Act and the Boulder Canyon Project Adjustment Act.* Washington, D.C.: Government Printing Office, 1947.

USBR. *Consumptive Uses and Losses Report.* Denver, Colo.: U.S. Bureau of Reclamation, 1996–. https://www.usbr.gov/uc/envdocs/plans.html.

USBR. *Final Environmental Impact Statement, Colorado River Interim Guidelines for Lower Basin Shortages and Coordinated Operations for Lake Powell and Lake Mead.* Boulder City, Nev.: U.S. Bureau of Reclamation, 2007.

USBR. *Colorado River Basin Water Supply and Demand Study.* Washington, D.C.: Government Printing Office, 2012.

U.S. Department of Commerce. *1982 Census of Agriculture: Volume 1, Part 6: Colorado State and County Data.* Washington, D.C: Bureau of the Census, 1984.

U.S. Department of the Interior. *Colorado River Accounting and Water Use Report: Arizona, California, and Nevada: Calendar Year 2017.* Boulder City, Nev.: Bureau of Reclamation, Boulder Canyon Operations Office, Lower Colorado Region, 2018. https://www.usbr.gov/lc/region/g4000/4200Rpts/DecreeRpt/2017/2017.pdf.

U.S. Department of the Interior. *Colorado River Basin Water Supply and Demand Study: Study Report.* Washington, D.C.: Bureau of Reclamation, 2012. https://www.usbr.gov/lc/region/programs/crbstudy/finalreport/Study%20Report/CRBS_Study_Report_FINAL.pdf.

U.S. Department of the Interior. *Compilation of Records in Accordance with Article V of the Decree of the Supreme Court of the United States in Arizona v. California, Dated March 9, 1964.* Boulder City, Nev.: Bureau of Reclamation, Boulder Canyon Operations Office, Lower Colorado Region, 2000. https://www.usbr.gov/lc/region/g4000/4200Rpts/DecreeRpt/2000DecreeRpt.pdf.

U.S. Department of the Interior. *Upper Colorado River Basin Consumptive Uses and Losses Report: 2011–2015*. February 2017. https://www.usbr.gov/uc/envdocs/reports/ColoradoRiverSystemConsumptiveUsesandLossesReports/20170200-ProvisionalUpperColoradoRiverBasin2011-2015-CULReport-UCRO.pdf.

USGS, *Surface Water Supply of the United States, 1930: Part IX, Colorado River Basin*. USGS Water Supply Paper 704. Washington, D.C.: Government Printing Office, 1931.

USGS, *Surface Water Supply of the United States, 1931: Part IX, Colorado River Basin*. USGS Water Supply Paper 719. Washington, D.C.: Government Printing Office, 1932.

U.S. Reclamation Service. *Problems of Imperial Valley and Vicinity*. S. Doc. 142, 67th Cong., 2nd sess. (1922).

Wheeler, George M. *Annual Report upon the Geographical Surveys West of the One Hundredth Meridian, in California, Nevada, Utah, Colorado, Wyoming, New Mexico, Arizona, and Montana: Being Appendix JJ of the Annual Report of the Chief of Engineers for 1876*. Washington, D.C.: Government Printing Office, 1876.

Wilbur, Ray Lyman, and Northcutt Ely. *The Hoover Dam Documents*. No. 717. Washington, D.C.: Government Printing Office, 1948.

Wilbur, Ray Lyman, and Northcutt Ely. *The Hoover Dam Power and Water Contracts and Related Data: With Introductory Notes*. Washington, D.C.: Government Printing Office, 1933.

Journal Articles, Newspaper Reports, Theses

Brean, Henry. "Drought Keeps Marinas on the Move." *Las Vegas Review-Journal*, February 7, 2004.

"Canyon Flood Peril Pictured." *Los Angeles Times*, October 22, 1923.

Carlson, John U., and Alan E. Boles Jr. "Contrary Views of the Law of the Colorado River: An Examination of Rivalries Between the Upper and Lower Basins." In *Proceedings of the Thirty-Second Annual Rocky Mountain Mineral Law Institute*, 21-1–21-68. New York: Bender, 1986.

Christensen, Niklas S., et al. "The Effects of Climate Change on the Hydrology and Water Resources of the Colorado River Basin." *Climatic Change* 62, no. 1–3 (2004): 337–63.

Clyde, Edward W. "Present Conflicts on the Colorado River." *Rocky Mountain Law Review* 32 (1960): 534–65.

Corthell, N. E. "The Colorado River Problem." *American Bar Association Journal* 9, no. 5 (1923): 289–92.

Corthell, N. E. "The Colorado River Problem—II." *American Bar Association Journal* 9, no. 6 (1923): 387–91.

Crawford, Arthur. "Sibert Report Jolts Boulder Dam Sponsors." *Chicago Tribune*, December 4, 1928.

Fleck, John. "What Seven States Can Agree to Do: Deal-Making on the Colorado River." Stanford University Rural West Initiative. Last modified August 3, 2012. http://web.stanford.edu/group/ruralwest/cgi-bin/drupal/content/what-seven -states-can-agree-do-deal-making-colorado-river.

Holdren, G. Chris, and Kent Turner. "Characteristics of Lake Mead, Arizona–Nevada." *Lake and Reservoir Management* 26, no. 4 (2010): 230–39.

Kelly, William. "The Colorado River Problem." *Transactions of the American Society of Civil Engineers* 88, no. 1 (1925): 306–47.

LaBianca, Margaret Bushman. "The Arizona Water Bank and the Law of the River." *Arizona Law Review* 40 (1998): 659–80.

Langbein, Walter B. "L'Affaire LaRue." *WRD Bulletin*, April–June 1977.

Leveen, E. Phillip. "Natural Resource Development and State Policy: Origins and Significance of the Crisis in Reclamation." *Antipode* 11, no. 2 (1979): 61–79.

MacDonnell, Lawrence J. "*Arizona v. California*: Its Meaning and Significance for the Colorado River and Beyond After Fifty Years." *Arizona Journal of Environmental Law and Policy* 4 (2013): 88–129.

MacDonnell, Lawrence J. "*Arizona v. California* Revisited." *Natural Resources Journal* 52, no. 2 (2012): 363–420.

McCabe, Gregory J., David M. Wolock, Gregory T. Pederson, Connie A. Woodhouse, and Stephanie McAfee. "Evidence That Recent Warming Is Reducing Upper Colorado River Flows." *Earth Interactions* 21, no. 10 (2017): 1–14.

McGee, William J. "Water as a Resource." *Annals of the American Academy of Political and Social Science* 33, no. 3 (1909): 37–50.

Meko, David M., et al. "Medieval Drought in the Upper Colorado River Basin." *Geophysical Research Letters* 34, no. 10 (2007).

Meyers, Charles J., and Richard L. Noble. "The Colorado River: The Treaty with Mexico." *Stanford Law Review* 19, no. 2 (1966): 367–419.

Midvale, Frank. "Prehistoric Irrigation in the Salt River Valley, Arizona." *Kiva* 34, no. 1 (1968): 28–32.

"Mighty Colorado Now Must Work for Man." *New York Times*, December 23, 1928.

"New Homes for 3,000,000 in Biggest Irrigation Project." *New York Times*, January 21, 1923.

Parsons, Malcolm Barningham. "The Colorado River in Arizona Politics." Master's thesis, University of Arizona, 1947.

Peterson, Thomas C., William M. Connolley, and John Fleck. "The Myth of the 1970s Global Cooling Scientific Consensus." *Bulletin of the American Meteorological Society* 89, no. 9 (2008): 1325–38.

Revelle, Roger, and Paul E. Waggoner. "Effects of a Carbon Dioxide-Induced Climatic Change on Water Supplies in the Western United States." In *Changing Climate: Report of the Carbon Dioxide Assessment Committee*. Washington, D.C.: National Academy Press, 1983.

Sarewitz, Daniel. "How Science Makes Environmental Controversies Worse." *Environmental Science and Policy* 7, no. 5 (2004): 385–403.

Smith, George Edson Philip. *A Discussion of Certain Colorado River Problems*. Tucson: University of Arizona, 1925.

Stockton, C. W., and Jacoby, Gordon C., Jr. "Long-Term Surface-Water Supply and Streamflow Trends in the Upper Colorado River Basin Based on Tree-Ring Analyses." *National Science Foundation Lake Powell Research Project Bulletin* 18 (1976): 1–70.

Trani, Eugene P. "Hubert Work and the Department of the Interior, 1923–28." *Pacific Northwest Quarterly* 61, no. 1 (1970): 31–40.

Tyler, Daniel. "Delphus Emory Carpenter and the Colorado River Compact of 1922." *University of Denver Water Law Review* 1 (1997): 228–74.

Udall, Bradley, and Jonathan Overpeck. "The Twenty-First Century Colorado River Hot Drought and Implications for the Future." *Water Resources Research* 53, no. 3 (2017): 2404–18.

Woodhouse, Connie A., et al. "Increasing Influence of Air Temperature on Upper Colorado River Streamflow." *Geophysical Research Letters* 43, no. 5 (2016): 2174–81.

Young, Robert A. "Coping with a Severe Sustained Drought on the Colorado River: Introduction and Overview." *Journal of the American Water Resources Association* 31, no. 5 (1995): 779–88.

INDEX

Note: Page numbers followed by "f" refer to figures; those followed by "t" represent tables. The word "Compact" in subentries refers to Colorado River Compact.

ABOUT THE AUTHORS

Eric Kuhn, now retired, worked for the Colorado River Water Conservation District from 1981 to 2018, including twenty-two years as general manager. The district is a water utility and policy agency covering most of the Colorado River Basin within Colorado.

John Fleck is director of the University of New Mexico's Water Resources Program. A Colorado River expert, he wrote *Water is for Fighting Over and Other Myths About Water in the West.*